Mark A. Young
1/9/95

T5-AEW-823

Restoration of Petroleum-Contaminated Aquifers

40000012
WALLACE'S
$90.50

Stephen M. Testa
Duane L. Winegardner

LEWIS PUBLISHERS

Library of Congress Cataloging-in-Publication Data

Testa, Stephen M.

    Restoration of petroleum-contaminated aquifers / by Stephen M.
Testa and Duane L. Winegardner.
       p. cm.
    Includes bibliographical references and index.
    ISBN 0-87371-335-4
    1. Oil pollution of water. 2. Water, Underground—Purification.
I. Winegardner, Duane L. II. Title.
TD427.P4T47   1991
628.1′6833—dc20
                                        90-13227
                                            CIP

COPYRIGHT © 1991 by LEWIS PUBLISHERS
ALL RIGHTS RESERVED

This book represents information obtained from authentic and highly regarded
sources. Reprinted material is quoted with permission, and sources are indicated.
A wide variety of references are listed. Every reasonable effort has been made
to give reliable data and information, but the author and the publisher cannot
assume responsibility for the validity of all materials or for the consequences of
their use.

Neither this book nor any part may be reproduced or transmitted in any form or
by any means, electronic or mechanical, including photocopying, microfilming,
and recording, or by any information storage and retrieval system, without
permission in writing from the publisher.

Direct all inquiries to CRC Press, Inc., 2000 Corporate Blvd., N.W., Boca Raton,
Florida 33431.

PRINTED IN THE UNITED STATES OF AMERICA
   3 4 5 6 7 8 9 0
Printed on acid-free paper

To my dear wife, Lydia, and my family for their continued support.

Stephen M. Testa

To my lovely wife, Jane.

Duane L. Winegardner

PREFACE

Over the past decade, the recognition of petroleum-contaminated groundwater has become widespread in the United States and internationally. Much attention has been focused upon uncontrolled releases of petroleum product being stored in underground storage tanks. These types of releases in numerous cases have resulted in adversely affecting groundwater quality and negatively impacting our groundwater resources. Uncontrolled releases from underground storage tanks, although ubiquitous, were generally localized in their vertical and lateral extent. In the late 1980s, the large-scale regional impact of petroleum products and their derivatives on groundwater resources was recognized. This reflected the uncontrolled and virtually unnoticed loss of hundreds of thousands of barrels of product being released into the subsurface, notably from refineries, but also from bulk-liquid above-ground storage terminals and pipeline corridors. As a result, a large amount of effort has been invested in the restoration of aquifers contaminated by petroleum-related operations. A large percentage of these sites are contaminated by spilled or leaked petroleum products. Some of the petroleum chemicals are free floating, some dissolved, and others are present as residual retained on the aquifer structures. Subsurface investigations at thousands of underground storage tank facilities plus many refineries, terminals, and pipelines have revealed a significant number of sites that will require remediation. Cleanup of these sites is an important factor in the protection and preservation of associated aquifers.

Environmental issues associated with such occurrences include hydrocarbon vapors, hydrocarbon contaminated soils, the presence of light nonaqueous phase liquid (LNAPL) hydrocarbon and dissolved hydrocarbon constituents in groundwater. The subsurface presence of hydrocarbons, notably the potential of LNAPL and dissolved constituents, are detected by the drilling of borings and subsequently installation of monitoring wells. Evaluation of the magnitude of the problem and

iv

ultimate remediation strategy depends upon the detection of LNAPL and volume evaluation thereof. Remediation strategies initiate with the delineation and recovery of the LNAPL, which serves as a continued source for groundwater contamination until removed. Once removed, dissolved hydrocarbon constituents in soils and groundwater are then addressed. Numerous recovery and aquifer restoration approaches are available and the prudent selection of an approach or combination of approaches requires a comprehensive understanding of the geologic and hydrogeologic environment, the subsurface occurrence of hydrocarbon, and site-specific factors such as waste-water handling capabilities and options. Monitoring the effectiveness and efficiency of such recovery and restoration operations allows for cost-effective system modifications and expansion, and enhanced production.

The objective of this book is to present the state of the knowledge on the restoration of aquifers contaminated by petroleum products and their derivatives. This book is intended for use by managers, regulators, consultants, and, notably, the practicing office and field remediation specialist. For those readers interested in advanced discussions, a comprehensive reference section follows each chapter, which includes a listing of pertinent papers of a developmental and theoretical nature. The book is organized in 11 chapters. An introduction to the subject matter is provided in Chapter 1. The regulatory environment and the framework that provides the mechanism in which environmental issues are addressed is discussed in Chapter 2. The geochemistry of petroleum and its fate and transport in the subsurface is presented in Chapters 3 and 4, respectively. Cleanup of groundwater in a petroleum-contaminated site can only be successful after the free-phase and residual product have been removed. Detection, occurrence, and behavior of immiscible light nonaqueous liquid hydrocarbons and subsequent recoverability is therefore fully addressed in Chapters 5 and 6, respectively. Particular attention is directed to solving problems associated with apparent thickness measurements, rising and falling water tables, trapped residual product, and appropriate pumping technologies. Chapter 7 provides various restoration methodologies and techniques for the recovery of immiscible LNAPL. Chapters 8 and 9 address the remediation of dissolved hydrocarbon constituents in groundwater and soil, respectively. Several procedures are discussed in Chapter 8 that are useful in recovering a wide variety of products with different physical and chemical properties. Cost concerns are discussed in Chapter 10. Lastly, actual case histories conducted by the authors and close associates are presented in Chapter 11. It is hoped that these studies serve as conceptual models and will aid the reader in the selection of the most appropriate technology and equipment available.

The authors gratefully acknowledge the contributions made to this book by M. T. Paczkowski of Gergherty & Miller in Philadelphia, Pennsylvania; Dr. E. L. Stover of Stover and Associates in Stillwater, Oklahoma; and M. M. Gates of

Gergherty & Miller in Tulsa, Oklahoma; for their contribution of Chapter 8; Dr. Duane R. Hampton of Western Michigan University, Michigan; for his discussion of empirical methods in evaluating actual vs. true immiscible LNAPL thicknesses as measured in monitoring wells. The review of certain sections was provided by several individuals, notably Keith Green of Applied Environmental Services, Ghulam Iqbal of Engineering Enterprises, Inc. General assistance was also provided by Jay Boughtner, Michael Contrell, Tom Danaher, Dawn Garcia, Dean Kirk, and Patrick L. Francks.

This work could not have been completed without the word processing and editorial assistance, including numerous other tasks, performed by Lydia Testa. Technical illustration assistance was provided by Mr. Tony Carmouche. To all we are very grateful.

Stephen M. Testa   Duane L. Winegardner
President   President
Applied Environmental Services   American Environmental Consultants
Laguna Hills, California   Norman, Oklahoma

ABOUT THE AUTHORS

Stephen M. Testa is President and founder of Applied Environmental Services located in Laguna Hills, California. Stephen received his B.S. and M.S. in Geology from California State University at Northridge, California. For the past 15 years, Stephen has worked in the areas of geology, hydrogeology, engineering geology, and hazardous waste management with firms such as Bechtel, Inc., Converse Consultants, Dames and Moore, and Engineering Enterprises, Inc.

Stephen has participated in numerous subsurface hydrogeologic site characterization projects associated with nuclear hydroelectric power plants, hazardous waste disposal facilities, and other industrial and commercial complexes. For the past 5 years, Stephen's main emphasis has been in the area of nonaqueous phase liquid hydrocarbon recovery and aquifer restoration. Maintaining overall management and technical responsibilities in engineering geology, hydrogeology, and hazardous waste related projects, he has participated in numerous projects involving oil recovery, hydrogeologic site assessments, soil contamination, water quality assessments, and the design and development of groundwater monitoring and aquifer remediation programs.

Stephen is the author of over 50 technical papers and 3 books, and is professionally active. He is a member of numerous organizations including the American Association for the Advancement of Science, Association of Engineering Geologists, American Association of Petroleum Geologists, Association of Groundwater Scientists and Engineers, Hazardous Materials Control Research Institute, Geological Society of America, California Groundwater Association, and Sigma Xi. He is also a member of the American Institute of

Professional Geologists where he has served on various committees, including the National Committee for Professional Development and Continuing Education. Stephen also conducts numerous workshops on various environmental aspects of the subsurface presence of petroleum hydrocarbons and teaches Hazardous Waste Management, Geology, and Mineralogy at California State University at Fullerton, California, and the University of Southern California.

Duane L. Winegardner is President of American Environmental Consultants, Inc. located in Norman, Oklahoma. He received his B.S. in Geology and M.S. in Geology and Hydrology from the University of Toledo, Toledo, Ohio. Subsequently, he has achieved registration as a Professional Engineer (Civil), and is currently licensed in several states. For the past 20 years, his work has focused on applied technology in the construction and environmental industries. His employers have included Toledo Testing Laboratory, St. Johns River Water Management District (FL), Environmental Science and Engineering, O.H. Materials Corporation, and Engineering Enterprises, Inc.

Since the advent of RCRA, Duane has been primarily active in investigation, design, and implementation of remediation at facilities with soil and groundwater contamination. Many of the remediation efforts were based on new applications of existing technology as well as the development of unique processes for specific geological and chemical settings. As technical manager (or design engineer), he has been responsible for the recovery, treatment, or containment of a wide variety of chemical products. During the most recent 4 years, his emphasis has been directed toward remediation of petroleum-contaminated sites.

Duane has published several technical papers and made presentations at numerous national seminars. He is an active member of ASTM and routinely participates in educational programs.

CONTENTS

1   INTRODUCTION

"Water is truly a mineral resource, dependent not merely upon the degree of usefulness, but upon the scarcity"

Never before has the interest in a healthy environment been such a strong stimulant to the development of a particular branch of scientific practice, as has been the case with groundwater science. Until recently, the occurrence of subsurface water was primarily studied to determine the most efficient procedure to recover it as a source of potable or irrigation water, or possibly to remove it from mines or construction sites. The traditional approach to aquifer evaluation has recognized the complexity of subsurface materials and conditions as a multidisciplinary science. Contributions to its development have come from geology, civil engineering, agricultural science, chemistry, and a variety of associated fields of interest.

When aquifer restoration began to be recognized as an important aspect of environmental management, the variety of scientific involvement increased even further. The change from a descriptive science (which identifies the existing subsurface characteristics and provides the engineering necessary to work with them) to a prescriptive science (which modifies the subsurface conditions in a controlled manner) requires a much better understanding of *all* the operating parameters. Physical, chemical, and biological processes are all interactive in a dynamic setting.

At specific sites where the contaminants are petroleum-related products, the spectrum of necessary professional expertise is greatly expanded. Recovery of multiphased petroleum products is the primary purpose of petroleum practice. Sorption of organic materials on soil grains has been studied in detail by agronomists. Biodegradation of organic chemicals is a microbiologist's dream. Procedures for determining concentrations or reactions of chemicals and their byproducts are within the scope of chemistry. The listing could be extended to

1

include many other scientific, engineering, or health science disciplines, the length of the list is without parallel.

The motivating force for the development of remediation science has been public opinion which has been expressed in the form of regulations. Almost every level of governing authority has enacted some form of environmental control. Interpretation of these rules is the realm of the attorney; therefore this group of professionals joins the scientific community as an important member of the remediation team.

This book has been written to acquaint the reader with the basic fundamentals of the more important topics related to this field. Some of the material will undoubtedly be a review to some readers; however, nearly all practicing professionals will find material that will be new to them or will serve to supplement their current knowledge.

Throughout the book, an effort has been made to limit the theoretical discussion to that required for basic understanding and to emphasize the practical applications. The inclusion of many case histories is intended to demonstrate technology that was appropriate for a particular site under the existing regulations. Before selecting or installing a remediation system, the designer is encouraged to evaluate all available site data, regulatory constraints, economic considerations, and safety precautions.

The organization of this book is intended to guide the reader toward the solution of real world problems. It will be helpful to read the entire book lightly, and then to concentrate on those sections of greater concern. The authors recognize that other procedures and interpretative opinions may be well suited for the solution of particular local problems. The discussion presented in these pages is representative of widely recognized and proven practices.

As a developing technology, innovative new procedures are being developed or borrowed from other technical fields. What is important is that the remediation effort is successful. The reader is invited to participate in the advancement of this profession.

2   REGULATORY FRAMEWORK

"Understanding of the pervasive regulations is essential to comprehension of the mechanism which drives and operates the environmental machine"

2.1   INTRODUCTION

Several sectors of the petroleum industry have been tightly regulated in certain environmental areas for almost two decades. This regulatory emphasis reflects a growing awareness of our limited available resources and thus has focused on those aspects. Notably, much regulatory attention has been aimed at the ubiquitous occurrence of accidental leaks and spills of hydrocarbon product from pipelines, underground storage tanks (UST), above-ground storage tanks, transport vehicles, and operational-related activities. Some of these tightly and rigorously regulated areas, as promulgated under the U. S. Environmental Protection Agency (EPA), include the Underground Storage Tank (UST) program, Underground Injection Control (UIC) program under the Safe Drinking Water Act (Title 40 Code of Federal Regulations [CFR] Part 144), and the National Pollutant Discharge Elimination System (NPDES) under the Clean Water Act (40 CFR Part 122). A summary of major federal and certain recent state regulations affecting petroleum hydrocarbons is presented in Table 2.1.

In addition, many states and local regions have developed and implemented more stringent regulations governing these programs, and subsequent remedial actions where hydrocarbons have been released to soil and groundwater. However, up to about December 1970 certain sectors of the petroleum industry, notably the exploration, development, and production activities and operations have remained essentially unregulated. For the past two decades, growing concerns regarding actual and anticipated groundwater problems in these areas have existed. These concerns reflect in part 150 years of operations and activities in the petroleum exploration, production, and refining industry. Although many

3

Table 2.1   Summary of Major Federal and Certain State Regulations Affecting Petroleum Hydrocarbons

Regulation	Description	Reference
Federal Water Pollution Control Act Amendments	Prohibits discharge of oil or hazardous substances from any vessel, from any onshore or offshore facility, into or upon the navigable waters of the U. S., adjoining shorelines, or the waters of the contiguous zone; also prohibits a discharge that would cause a visible sheen upon the water or adjoining shorelines or cause a sludge or emulsion to be deposited beneath the surface of the water	40 CFR Part 112 40 CFR Part 154 40 CFR Part 155 40 CFR Part 156 40 CFR Part 557
Federal Clean Water Act	Protects surface and groundwater quality to maintain "beneficial uses" of water	40 CFR Part 116 40 CFR Part 117 40 CFR Part 122
Resource Conservation and Recovery Act (RCRA)	Provides cradle-to-grave management of hazardous waste including regulation of hazardous waste generators, transporters, and treatment, storage and disposal facilities	40 CFR Parts 260–268, 270–272, 280–281
Safe Drinking Water Act (SDWA)	Protects drinking water sources from toxic contamination	40 CFR Part 124 40 CFR Part 144 40 CFR Part 145 40 CFR Part 146 40 CFR Part 147

Regulation	Description	Citation
Comprehensive Environmental Response, Compensation, and Liability Act (CERCLA)	Locates, assesses, and cleans up contaminated sites; also requires reporting of releases of hazardous chemicals	40 CFR Part 300
Department of Transportation	Provides for safe transportation of hazardous materials and wastes	40 CFR Part 195
Pipeline Safety Act of 1987	Provides for safe transportation of hazardous liquids	49 USC 2001 / 49 USC 2002
California Regional Water Quality Control Board (CRWQCB)	Requires characterization of subsurface hydrogeological conditions, delineation and chemical characterization of LNAPL product pools; recovery of LNAPL product and overall aquifer restoration and soil remediation	CRWQCB Order No. 85-17
Above-Ground Petroleum Storage Act	Provides requirements for establishing and maintaining a monitoring program; installation and maintenance of a leak detection system if potential exists to adversely impact surface water and groundwater	California Senate Bill No. 1050

favorable changes have come about in the way these operations and activities are conducted today, in recent years increasing environmental regulations are being felt throughout the industry. The industry has been fortunate to date in that it has temporarily escaped the designation and regulation of its wastes as hazardous, but with concerns about future clean-up costs, an increase in the level of regulatory attention to the conduct of oil and gas exploration, production, and refinery operations is anticipated.

Southern California has a rich history of oil and gas exploration and production going back to 1876, the first year of commercial production. Thirty out of 56 counties in California are known to produce oil and gas. Generally, south of 37° North latitude, crude oil occurs primarily in four basins: Los Angeles Basin, Ventura Basin, San Joaquin Basin, and Santa Maria Basin, as shown in Figure 2.1. The price of real estate is also at a premium in areas such as Los Angeles County where much of the available land that remains to be developed is oil-field property. These oil-field areas are rapidly undergoing environmental pressures as they reach the end of their productive lives. In addition, the conversion of manufacturing-related land to services-related land use and significant increases in land value has resulted in an increasing number of property transfers involving oil-field production and storage areas.

2.2   RESOURCE CONSERVATION AND RECOVERY ACT/COMPREHENSIVE ENVIRONMENTAL RESPONSE, COMPENSATION, AND LIABILITY ACT

The 1984 amendments to the Resource Conservation and Recovery Act (RCRA) required the EPA to develop federal regulations dealing with leakage of hydrocarbon product or substances from underground storage tanks designated as hazardous, as defined under the Comprehensive Environmental Response, Compensation, and Liability Act (CERCLA) of 1980 (Section 101[14]). In California, the regulatory responsibilities and potential liability of operators and owners of UST are set forth primarily in the California UST Law (Health and Safety Code Section 25280), California UST Regulations (California Code of Regulations, Subchapter 16, Title 23), and numerous local codes and ordinances. In the case of the UST program, the basic requirements of tank owners under the California Underground Storage Tank Regulations (CUSTR) provide provisions for existing tanks including:

1. Determine whether past leaks have occurred from the tank system;
2. Determine whether the tank system is currently leaking;
3. Implement a leak detection and monitoring program to detect future releases.

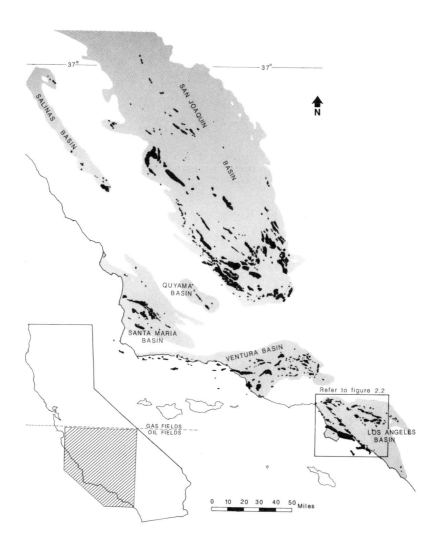

Figure 2.1.   Location of major basins and oil fields in southern California.

However, as straightforward as this may seem, there are 23 local health and fire departments which have adopted their own version of this bill, making compliance increasingly difficult.

The historical usage of USTs for the bulk storage of petroleum products was instigated by the need to store flammable petroleum products underground to reduce fire risk and to increase the financial savings in respect to space which could be utilized for other purposes by storing large volumes of petroleum underground. The use of USTs for petroleum storage made excellent sense; however, in the early 1980s petroleum leakage problems of enormous magnitude were being reported by the general public all across the United States. The problems resulted from leakage of petroleum products from USTs. These releases of stored liquids had contaminated both soil and groundwater. Some releases had contaminated nearby private and municipal drinking water supply wells. The release of petroleum products into soils created problems aside from the obvious soil contamination. Explosive concentrations of gasoline vapors could migrate through porous soils and accumulate in sewer and utility conduits as well as in the basements of buildings. These vapors could potentially pose a substantial fire and explosion hazard. In one case in Colorado in 1980, 41 homes were purchased at over twice their appraised value at a cost of ten million dollars to the responsible party as a result of an UST leak. In another instance, five to ten million dollars as of May 1984, was spent for clean-up and damages resulting from 30,000 gallons of gasoline that leaked from an UST into a New York community groundwater supply in 1978.

Following the report of the U. S. General Accounting Office in 1984 that UST leaks had been reported in all 50 states, many states in 1985 claimed that USTs were the leading cause of underground contamination. As of 1984, there were an estimated 2.5 million underground storage tanks in the United States of which about 97% are used to store petroleum products. By 1986, the Environmental Protection Agency (EPA) identified 12,444 incidents of release that had occurred as of 1984.

A comprehensive study was undertaken by the EPA in 1987 to gather additional facts about releases from USTs. This study resulted in the following findings:

- Most releases do not originate from the tank portion of UST systems, but pipeline releases occur twice as often as tank releases;
- Spills and overfills during refueling are the most common cause of releases;
- Older steel tanks fail primarily because of corrosion but new cathodically protected steel, fiberglass-clad steel, and fiberglass tanks have nearly eliminated external corrosion failure;
- Corrosion, poor installation, accidents, and natural events are the four major causes of piping failure; and
- When pressurized piping fails, significantly larger releases can occur.

This study was very important in the formulation of federal UST regulations. UST systems are located at over 700,000 facilities nationwide. Over 75% of the existing systems are made of unprotected steel, a type of tank shown by studies to be the most likely type to leak. Most of these facilities are owned and operated by small "Mom & Pop" enterprises that do not have easy access to the financial resources for potential clean-up costs and are not accustomed to dealing with complex regulatory requirements. Current innovations and technologies associated with USTs are changing rapidly and keeping abreast of them is sometimes difficult and immensely time consuming. In response to these unique aspects associated with USTs, the EPA identified several key objectives that it would have to incorporate into the regulations to be formulated for USTs. These objectives were

1. The regulatory program must be based upon sound national standards that protect human health and the environment.
2. The regulatory program must be designed to be implemented at a state and local level.
3. The regulations must be kept simple and easily understood and implemented.
4. The regulations must not impede technological developments.
5. The regulations must retain some degree of flexibility.

With these key objectives in mind, the EPA formulated the Federal UST Regulations, as described in 40 CFR, Parts 280 and 281, effective December 22, 1988, with financial responsibility requirements effective January 24, 1988. It should also be noted that fire safety matters are not emphasized in the EPA UST regulations. The National Fire Protection Association Code 30 (NFPA Code 30), relating to the storage and handling of flammable and combustible liquids, is used by the majority of local regulatory communities in the United States in regulating fire safety tanks that are used to store petroleum products.

On November 8, 1984, President Reagan signed into law the Hazardous and Solid Waste Amendment Act of 1984 (HSWA). Subtitle I of HSWA amended the Resource Conservation and Recovery Act (RCRA) of 1976 to specifically address the regulation of UST. Under section 9003 of RCRA, the EPA had to establish regulatory requirements for the following:

- Exemptions
- UST systems design, construction, installation, and notification
- General operating requirements
- Release detection
- Release prevention
- Release reporting, investigation, and confirmation
- Release resource and corrective action
- UST closures
- State program approval
- Enforcement and penalties
- Financial responsibility

Standards for financial responsibility were promulgated by the Superfund Amendment and Reauthorization Act of 1986 (SARA), which further amended Section 9003 of RCRA and mandated that the EPA establish financial responsibility requirements for UST owners and operators so as to guarantee cost recovery for corrective action and third party liability caused by accidental releases of UST containing petroleum products.

2.3   DEPARTMENT OF TRANSPORTATION

In addition to the statistics developed for UST, in 1984 the Office of Technology Assessment also reported that 16,000 spills occur annually during transport, while the Department of Transportation (DOT) reported that of the 4112 accidents that occurred between 1968 and 1981, 1372 were associated with corrosion and 1101 with pipeline ruptures. Prior to DOT regulating the transport of liquids by pipeline in 1971, 308 interstate pipeline accidents were documented, resulting in a loss of about 245,000 barrels of liquid. Increasing DOT involvement may have been the reason for reduced number of incidents in 1980 (275), and 1981 (198). The DOT came into being in 1967 and was charged with the responsibility of pipeline safety. The safety code, made effective in 1970, drew extensively on the voluntary code of the industry. Standards concerned with the design and operation of pipelines that were incorporated into the regulations reflected the composite experience of individuals, companies, and professional societies over several years. These efforts were supplemented by research, tests, studies, and investigations. In 1976, DOT reported that 83% of pipeline accidents were traceable to the line pipe with other accidents traceable to tank farms, pumping stations, and delivery points with successively decreasing percentages. However, the regulations emphasized safety with regard to injuries, not necessarily contamination.

Pipeline safety is currently promulgated by the DOT Office of Pipeline Safety under the Natural Gas Pipeline Safety Act of 1968 and the Hazardous Materials Transportation Act as summarized in Table 2.2.

Petroleum releases are addressed under 40 CFR Part 195, which is subdivided into six subparts (Subparts A through F). Although these regulations apply to pipeline facilities and the transportation of hazardous liquids associated with those facilities (in or affecting interstate or foreign commerce), they do not apply to transportation of hazardous liquids via the following:

• Gaseous state
• Pipeline by gravity
• Pipeline stress level of 20% or less of the specified minimum yield strength of the live pipe

Table 2.2   Department of Transportation Code of Federal Regulations

Code of Federal Regulations Title	Part Description
40 CFR Part 190	Pipeline Safety Program Procedures
40 CFR Part 191	Transportation of Natural and Other Gas by Pipeline; Annual Reports and Incident Reports
40 CFR Part 192	Transportation of Natural and Other Gas by Pipeline: Minimum Federal Safety Standards
40 CFR Part 193	Liquified Natural Gas Facilities: Federal Safety Standards
40 CFR Part 195	Transportation of Hazardous Liquids by Pipeline

- Transportation in onshore gatherlines in rural areas
- Transportation in offshore pipelines located upstream from the outlet flange of each facility on the outer continental shelf where hydrocarbons are produced or where produced hydrocarbons are first separated, dehydrated, or processed (whichever facility is farther downstream)
- Transportation via onshore production (including flow lines), refining, or manufacturing facilities, or storage or in-plant piping systems associated with such facilities
- Transportation by vessel, aircraft, tank truck, tank car, or other vehicle or terminal facilities used exclusively to transfer hazardous liquids between such modes of transportation

The Pipeline Safety Act of 1987 provides regulations establishing federal safety standards, including requirements for release detection, prevention, and correction, for the transportation of hazardous liquids and pipeline facilities, as necessary to protect human health and the environment. These regulations apply to each person who engages in the transportation of hazardous liquids or who owns or operates pipeline facilities.

Of notable importance is the lack of regulations pertaining to pipelines associated with refining or manufacturing facilities, or associated storage or in-plant piping systems that are landbound and are situated above aquifers of beneficial use.

2.4   CLEAN WATER ACT

Spills of petroleum products continue to be regulated primarily under Section 311 of the Clean Water Act. Spills at deepwater ports or from outer continental shelf oil and gas operations are also governed in part under the Deepwater Ports Act of 1974 and Outer Continental Shelf Act Amendments of 1978. Although oil and hazardous substances are covered under the same Clean Water Act provision traditionally, they have been treated separately by the EPA.

2.4.1 Oil Spills

Regarding oil spills, owners and operators with large oil storage facilities (1320 gallons above ground or 42,000 gallons below ground) must comply with 40 CFR Part 112. This title requires the development, implementation, and maintenance of Spill Prevention, Control, and Countermeasure (SPCC) plans. SPCC plans typically include provisions for the installation of containment structures, regular inspections, and other preventative measures.

More stringent vessel safety and accident and spill prevention measures are regulated by the Coast Guard, as promulgated under the Port and Tanker Safety Act of 1978, 33 United States Code, Sections 1201 to 1231, Safety of Life at Sea (SOLAS) Convention of 1973, and the MARPOL Convention of 1978. Financial responsibility to meet potential liability under the Clean Water Act must also be met.

2.4.2 Hazardous Substance Spills

Section 311 of the Clean Water Act also governs the discharge of hazardous substances. Approximately 300 substances were designated as hazardous (40 CFR Part 116). In addition, EPA designated quantities of these substances that may be considered harmful (i.e., reportable quantity) (40 CFR Part 117). Five categories (X, A, B, C, and D) were designated wherein:

- X substance is 1 pound
- A substance is 10 pounds
- B substance is 100 pounds
- C substance is 1000 pounds
- D substance is 5000 pounds

Discharges made in compliance with an NPDES permit are excluded, reflecting the purpose of the amendments to limit "classic" hazardous substance spills, not chronic discharge of designated substances if the discharge complies with an NPDES permit. In addition, the facility has the option to regulate intermittent anticipated spills (i.e., into plant drainage ditches) under its existing NPDES permit. Discharges from industrial facilities to a Publicly Owned Treatment Works (POTW) are not currently regulated. However, the regulations can apply to all discharges of reportable quantities of hazardous substances to POTWs by a mobile source such as trucks under certain circumstances (40 CFR Part 117). Reporting requirements for spills have been significantly supplemented by CERCLA and Title III of the Superfund Amendments and Reauthorization Act (SARA) of 1986.

2.5   SAFE DRINKING WATER ACT

The Safe Drinking Water Act (SDWA) of 1974 and the amendments made to it in 1980 mandated that primacy states regulate injection wells to protect underground sources of drinking water. Underground Sources of Drinking Water (USDW) are defined as aquifers or portions of aquifers with a total dissolved solids value of less than 10,000 mg/l and that are capable of supplying a public water system. Approximately 60 million barrels of oil-field fluids were injected through 166,000 injection wells within the conterminous United States in 1986. These volumes are anticipated to significantly increase in the future as producing fields continue to be depleted. Thus, construction requirements as listed in 40 CFR 146.22 are an essential prerequisite to the safe disposal/injection of fluids and the prevention of contamination of USDW.

There are five classes of injection wells (Class I through V), with Class II wells defined as those wells used in conjunction with oil and gas production activities. Class II wells are defined as those wells that inject fluids

- That are brought to the surface in connection with conventional oil or natural gas production and may be commingled with waste waters from gas plants that are an integral part of production operations, unless those waters are classified as a hazardous waste at the time of injection
- For enhanced recovery of oil or natural gas
- For storage of hydrocarbons that are liquid at standard temperature and pressure

2.6   CALIFORNIA REGIONAL WATER QUALITY CONTROL BOARD ORDER NO. 85-17

Several refineries, tank farms, and other petroleum-handling facilities are included on the EPA National Priorities List (EPA, 1986) or are being regulated under RCRA. Such facilities present several potential subsurface environmental concerns reflecting approximately 80 years of operation. These concerns include

- Soils containing elevated hydrocarbon concentrations such that the soil may be considered a hazardous waste
- LNAPL pools that have leaked from reservoirs, tank farms, and pipelines and serve as a source of soil contamination and a continued source for groundwater contamination
- Dissolved hydrocarbons in groundwater that may adversely affect water-bearing zones considered of beneficial use or as drinking water supplies
- Accumulation of vapors that could pose a fire or explosion hazard

Southern California prior to the turn of the century was composed of large Spanish land grants used primarily for agricultural purposes. With the discovery of oil, numerous refineries were constructed in the early 1920s in close proximity to both production areas as well as the shipping facilities of the Los Angeles Harbor. The majority of these refineries continued to expand their operations and areal extent through to the late 1940s. The locations of major oil fields throughout the Los Angeles coastal plain are shown in Figure 2.2. The locations of major refineries, tank farms, and pipeline corridors within the Los Angeles coastal plain are shown in Figure 2.3. Although many of these facilities were initially isolated and located in moderately to heavily industrialized areas, with the encroachment of development many are now in close proximity to densely populated light commercial and residential zones.

In early 1985, oil droplets were evident on beach sands near high-priced beachfront real estate just west of a refinery. The product evidently leaked from the nearby refinery situated near the waterfront, migrated downward to the shallow water table, and then moved laterally with the regional groundwater flow direction toward the ocean. Explosive concentrations of vapors were also detected at the bottom of a construction pit being excavated near the refinery and several nearby homes. A massive pollution problem was then recognized by the authorities. At this time, an estimated 6,000,000 barrels of product were thought to exist beneath the majority of the 1000 acre 80-year-old refinery.

The highly visible presence and potential hazard to public health, safety, and welfare prompted a minimum of 17 oil refineries and tank farms to be designated as health hazards by the California Department of Health Services (DOHS). This designation reflected the potential and actual subsurface occurrence of leaked hydrocarbon product derived from such facilities during 60 years of operations that has migrated through the subsurface, resulting in the presence of LNAPL pools overlying the water table. Although several of these refineries were listed as hazardous waste sites with remediation being required under RCRA, the majority of such facilities fell under the Los Angeles Region of the California Regional Water Quality Control Board Order No. 85-17 adopted in February 1985 for assessment and subsequent remediation. This order was the first to address large-scale regional subsurface environmental impacts by the petroleum refinery industry. This order required, in part, assessment of the subsurface presence of hydrocarbons and other associated groundwater pollutants that may

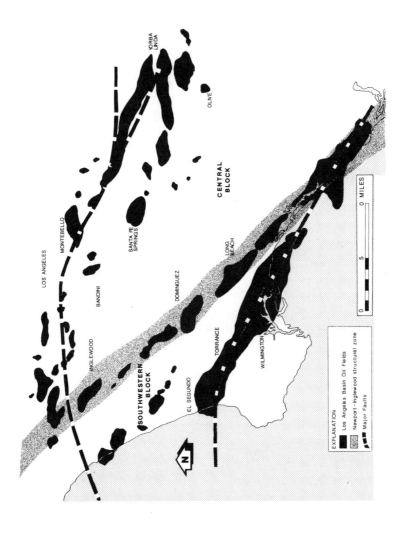

Figure 2.2.   Location of major oil fields in the Los Angeles Basin.

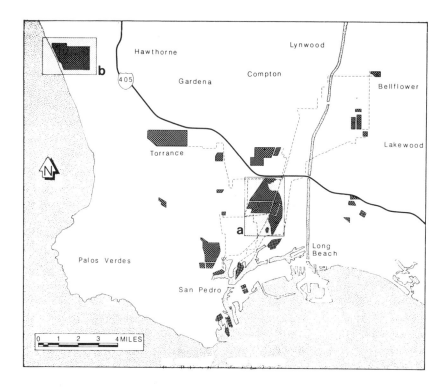

Figure 2.3.   **Locations of major petroleum handling facilities and structures, including refineries, above-ground storage tank farms,
and pipeline corridors in the Los Angeles coastal plain.**

affect subsurface soils and groundwater beneath such facilities. Specifically, the
following items were included:

- Characterization of subsurface geologic and hydrogeologic conditions
- Delineation of LNAPL pools including chemical characterization, and
  real extent and volume
- Implementation of LNAPL recovery
- Overall aquifer restoration and rehabilitation of dissolved hydrocarbons
  and associated contaminants
- Eventual soil remediation of residual hydrocarbons

The spatial distribution and areal extent of leaked LNAPL beneath the Los
Angeles coastal plain is shown in Figures 2.3a and 2.3b (Testa, 1990). The areal
extent of these LNAPL pools combined is on the order of 1000 acres. An
estimated minimum volume is on the order of approximately 1.5 million barrels
(ranging up to an estimated 7 million barrels) of product leaked from refineries
and major tank farms situated on the coastal plain.

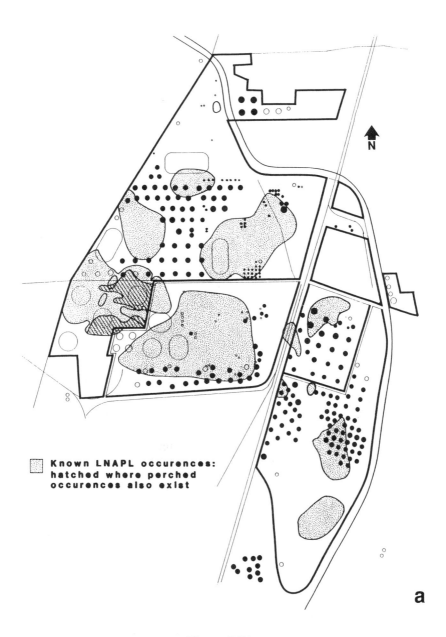

Figure 2.3A.

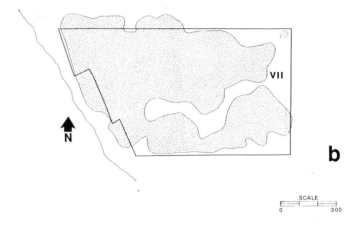

Figure 2.3B.

2.7   CALIFORNIA ABOVE-GROUND PETROLEUM STORAGE ACT

Regulations exist that prohibit the ownership or operation of an UST used for the storage of hazardous substances unless a permit for its operation is issued, in addition to imposing various design, installation, monitoring, and release reporting requirements for UST. However, there are many above-ground storage tanks presently in service that were constructed 20 or more years ago. This has allowed sufficient time for metal corrosion and stress-fatigue cracking to weaken and perhaps to breach the tank floors. The resultant tank floor leaks often go unnoticed by facility personnel during the performance of their daily routines. These leaks can be costly, both in terms of product loss and the potential liability for environmental pollution.

Until recently, techniques for tank-floor integrity testing were generally limited to visual inspection or still gauging. Visual inspections required the tank to be emptied, cleaned, and carefully examined, putting the tank out of service for weeks or more. Even then, some small leaks could be overlooked.

Still-gauging methods are adequate for only the largest leaks. The accuracy of most tank-installed liquid level gauges is usually 0.125 in. at best. A product loss reflected by a 0.062 in. level drop for a 100 ft. diameter tank translates to a 306 gal/day leak. If this 0.0625 in. drop in product level is not discernible from the masking effects of fluid expansion, losses in excess of 2,650 barrels annually will go undetected. For example, at $20 per barrel, this loss amounts to over $53,000 for a single tank. Most importantly, the associated liability risks of groundwater contamination could involve much greater potential costs.

In California, regulations are being proposed (Senate Bill No. 1050, Aboveground Petroleum Storage Act) to require each owner or operator of an above-ground storage tank at a facility to establish and maintain a monitoring program within 180 d of preparing a spill prevention control and countermeasure plan. Secondly, if the tank facility has the potential to impact surface waters or sensitive ecosystems as determined by the reviewing agency, based on the tank location, tank size, characteristics of the petroleum being stored, or the spill containment system, the facility owner will need to perform tasks similar to that of UST owners. These requirements would include either of the following:

- Install and maintain a system to detect releases into surface waters or sensitive ecosystems.
- If any discharge from a tank facility flows — or would reasonably be expected to flow — to surface waters or a sensitive ecosystem, allow for a drainage valve to be opened and to remain open only during the presence of an individual who visually observes the discharge.

Thirdly, if because of tank facility location, tank size, or characteristics of the crude oil or its fractions being stored (16 degrees API or lighter), a facility that has the potential to impact the beneficial uses of the groundwater, and that is not required to have a groundwater monitoring program at the tank facility pursuant to any other federal, state, or local law shall do any of the following:

- Install a tank facility groundwater monitoring system that detects releases of crude oil or its fractions into the groundwater
- Install and maintain a tank foundation design that will provide for early detection of releases of crude oil or its fractions before reaching the groundwater
- Implement a tank water bottom monitoring system and maintain a schedule that includes a log or other record that will identify or indicate releases of crude oil or its fractions before reaching the groundwater
- Use other methods that will detect releases of crude oil or its fractions into or before reaching the groundwater

All positive findings from the detection system, excluding tanks whose exterior surface (including connecting piping and the floor directly beneath the tank) can be monitored by direct viewing, must be reported to the appropriate agency within 72 hours after learning of the finding.

Federal legislation for above-ground storage tanks is also in the process of being mandated under RCRA. This bill would include requirements for development and implementation of a release prevention plan, a tank system that is capable of containing 110% of the tank content and preventing offsite release, and inspection of tank systems by a qualified professional engineer. Provisions will also include evidence of financial responsibility and clean-up of product releases. There is little doubt that legislation for above-ground storage tanks will be more stringent in the years to come.

REFERENCES

1. *Special Report: State Underground Petroleum Products Storage Mandates* (Washington, D.C., American Petroleum Institute, State Relations Department, September 21, 1984).
2. Arbuckle, J. G., et al., *Environmental Law Handbook*, 10th ed., (Rockville, MD) Government Institutes, Inc., 664 p.
3. Arscott, R. L., 1989, New Directions in Environmental Protection in Oil and Gas Operations, in *Environmental Concerns in the Petroleum Industry* (edited by Testa, S. M.), Pacific Section American Association of Petroleum Geologists Symposium Volume, p. 217-227.
4. Artz, N. S. and Metzler, S. C., 1985, Losses of Stored Waste Oil from Below-Ground Tanks and the Potential for Groundwater Contamination, in *Proceedings of the National Conference on Hazardous Wastes and Environmental Emergencies,* May, 1985, p. 60-65.
5. Buonocore, P. E., Ketas, G. F., and Garrahan, P. E., 1986, New Requirements for Underground Storage Tanks: in *Proceedings of the National Conference on Hazardous Wastes and Hazardous Materials,* March 4-6, 1986, p. 246-250.
6. California Code of Regulations, Title 23, Chapter 3, Subchapter 16, Article 4.
7. Dezfulian, H., 1988, Site of an Oil-Producing Property: *Proceedings of the Second International Conference on Case Histories in Geotechnical Engineering,* Vol. 1, p. 43-49.
8. Eger, C. K. and Vargo, J. S., 1989, Prevention: Ground Water Contamination at the Martha Oil Field, Lawrence and Johnson Counties, Kentucky: in *Environmental Concerns in the Petroleum Industry* (edited by Testa, S. M.), Pacific Section of the American Association of Petroleum Geologists Symposium Volume, p. 83-105.
9. Garcia, D. H. and Henry, E. C., 1989, Environmental Considerations for Real Estate Development of Oil Well Drilling Properties in California: in *Environmental Concerns in the Petroleum Industry* (edited by Testa, S. M.), Pacific Section American Association of Petroleum Geologists Symposium Volume, p. 117-127.
10. Government Institutes, Inc., 1989, *Environmental Statutes:* Government Institutes, Inc., Rockville, MD, 1169 p.
11. Hansen, P., 1985, L.U.S.T.: Leaking Underground Storage Tanks: in *Proceedings of the National Conference on Hazardous Waste and Environmental Emergencies,* May 14-16, 1985, p. 66-67.
12. Henderson, T., 1989, Assessment of Risk to Ground Water Quality from Petroleum Product Spills: in *Proceedings of the National Water Well Association and American Petroleum Institute Conference on Petroleum Hydrocarbons and Organic Chemicals in Ground Water: Prevention, Detection and Restoration,* NWWA, Houston, Texas, p. 333-345.

13. Jones, S. C. and O'Toole, P., 1989, Increasing Environmental Regulation of Oil and Gas Operations: in *Environmental Concerns in the Petroleum Industry* (edited by Testa, S. M.), Pacific Section American Association of Petroleum Geologists Symposium Volume, p. 209-215.

14. Lovegreen, J. R., 1989, Environmental Concerns in Oil-Field Areas During Property Transfers: in *Environmental Concerns in the Petroleum Industry* (edited by Testa, S. M.), Pacific Section American Association of Petroleum Geologists Symposium Volume, p. 129-158.

15. Meyer, C. F., 1973, *Polluting Ground Water: Some Causes, Effects, Controls and Monitoring:* U.S. EPA Environmental Monitoring Series, EPA-600/4-73-001b.

16. Michie, T., 1988, Oil and Gas Industry Water Injection Well Corrosion Study: *Proceedings of the UIPC Summer Meetings,* p. 47-67.

17. Mulligan, W. S., 1988, Waste Minimization in the Petroleum Industry: in *Proceedings of the Fourth Annual Hazardous Materials Management Conference West*, November 8-10, 1988, p. 540-543.

18. Office of Technology Assessment, 1984, *Protecting the Nation's Ground Water from Contamination - Volumes I and II:* OTA, Washington, D.C., No. OTA-0233, October, 503p.

19. Patrick, R., Ford, E. and Quarles, J., 1988, *Groundwater Contamination in the United States*: University of Pennsylvania Press, Second Edition, Philadelphia, 513 p.

20. Rowley, K., 1986, The Rules of the Games in Ground-Water Monitoring: in *Proceedings of the Second Annual Hazardous Materials Management Conference West*, December 3-5, 1986, p. 365-374.

21. Savini, J. and Kammerer, J. C., 1961, *Urban Growth and the Water Regimen*: United States Geological Survey Water-Supply Paper, No. 1591-A, 43 p.

22. Sittig, M., 1978, *Petroleum Transportation and Production - Oil Spill and Pollution Control*: Noyes Data Corporation, New Jersey, 360 p.

23. Syed, T., 1989, An Overview of the Underground Injection Control Regulations for Class II (Oil and Gas Associated) Injection Wells - Past, Present and Future: in *Environmental Concerns in the Petroleum Industry* (edited by Testa, S. M.), Pacific Section American Association of Petroleum Geologists Symposium Volume, p. 199-207.

24. Testa, S. M., 1989, Regional Hydrogeologic Setting and its Role in Developing Aquifer Remediation Strategies: in *Proceedings of the Geological Society of America*, 1989 Annual Meeting Abstracts with Programs, Vol. 21, No. 6, p. A96.

25. Testa, S. M., 1990, Light Non-Aqueous Phase Liquid Hydrocarbon Occurrence and Remediation Strategy, Los Angeles Coastal Plain, California: in *Proceedings of the International Association of Hydrogeologists, Canadian National Chapter, on Subsurface Contamination by Immiscible Fluids,* April, 1990, in press.

26. Testa, S. M., Henry, E. C. and Hayes, D., 1988, Impact of the Newport-Inglewood Structural Zone on Hydrogeologic Mitigation Efforts - Los Angeles Basin, California: *Proceedings of the National Water Well Association of Ground Water Scientists and Engineers FOCUS Conference on Southwestern Ground Water Issues,* p. 181-203.

27. Testa, S. M. and Townsend, D. S., 1990, Environmental Site Assessments in Conjunction with Redevelopment of Oil-Field Properties within the California Regulatory Framework: in *Proceedings of the National Water Well Association of Groundwater Scientists and Engineers Cluster of Conferences.*

28. U.S. Accounting Office, 1984, *Federal and State Efforts to Protect Groundwater:* Washington, D.C., February, 80 p.

29. U.S. Environmental Protection Agency, 1985, *National Water Quality Inventory 1984 National Report to Congress:* Office of Water Regulations and Standards, Washington, D.C., EPA-440/4-85-029, 173 p.

30. U.S. Environmental Protection Agency, 1986, Amendment to National Oil and Hazardous Substances Contingency Plan National Priorities List, Final Rule and Proposed Rule: Federal Register, v. 51, no. 111, June 10, p. 21053-21112.

31. U.S. Environmental Protection Agency, 1986, *Summary of State Reports on Releases from Underground Storage Tanks:* Office of Solid Waste, Washington, D.C., EPA-600/M-86-020, July, 95 p.

32. U.S. Environmental Protection Agency, Code of Federal Regulations, 1985, Title 40, Parts 124, 144, 145, 146, and 147.

33. U.S. Environmental Protection Agency, Code of Federal Regulations, 1984, Title 40, Parts 280 and 281.

34. U.S. Environmental Protection Agency, Code of Federal Regulations, 1980, Title 40, Section 101(14).

35. U.S. Environmental Protection Agency, Code of Federal Regulations, 1968, Title 40, Parts 190, 191, 192, 193, and 194.

36. U.S. Environmental Protection Agency, Code of Federal Regulations, 1985, Title 40, Parts 124, 144, 145, 146, and 147.

37. Wolbert, G. S., 1979, *U.S. Oil Pipe Lines - An Examination of How Oil Pipe Lines Operate and the Current Public Policy Issues Concerning Their Ownership:* American Petroleum Institute, Washington, D.C., 556 p.

3 GEOCHEMISTRY OF PETROLEUM HYDROCARBONS

"All substances are poison; there is none which is not a poison. The right dose differentiates a poison from a remedy" (Paracelcus, 1493-1541)

3.1 CHEMISTRY OF PETROLEUM

The term *petroleum* is derived from the Latin derivative petra for rock and oleum for oil. Current usage defines petroleum as any hydrocarbon mixture of natural gas, condensate, and crude oil. Most important to this discussion is crude oil which is a heterogenous liquid consisting of hydrocarbons comprised almost entirely of the elements hydrogen and carbon in a ratio of about 1.85 hydrogen atoms to 1 carbon atom. Minor constituents typically comprising less than 3% in total volume include sulfur, nitrogen, and oxygen. Trace constituents typically comprising less than 1% in total volume include phosphorus and heavy metals such as vanadium and nickel.

The composition may vary with the location and age of an oil field, and may even be depth dependent within an individual well. Crudes are commonly classified according to their respective distillation residue, which reflects the relative contents of three basic hydrocarbon structural types: paraffins, naphthenes, and aromatics. About 85% of all crude oils can be classified as either asphalt base, paraffin base or mixed base. Asphalt base crudes contain little paraffin wax and an asphaltic residue (predominantly condensed aromatics). Sulfur, oxygen, and nitrogen contents are often relatively higher in comparison with paraffin base crudes, which contain little to no asphaltic materials. Mixed-base crude contains considerable amounts of both wax and asphalt. Representative crude

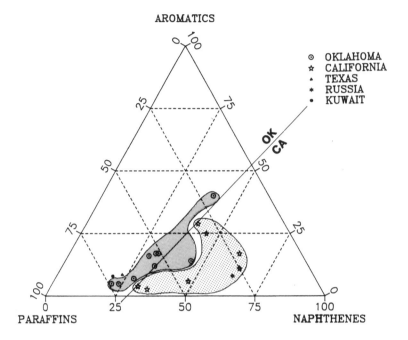

Figure 3.1. Ternary diagram showing representative crude oils and their respective composition with respect to paraffins, naphthenes, and aromatics. Note the clustering of the California crudes toward the napthenes and the Oklahoma crudes toward the paraffins.

oils and their respective composition in respect to paraffins, naphthenes, and aromatics are shown in Figure 3.1.

A petrochemical is a chemical compound or element recovered from petroleum or natural gas, or derived in whole or in part from petroleum or natural gas hydrocarbons, and intended for chemical markets. Petrochemicals or hydrocarbons in general are simply compounds of hydrogen and carbon that can be characterized based on their respective chemical composition and structure. Each carbon atom can essentially bond with four hydrogen atoms. Methane is the simplest hydrocarbon and is illustrated as follows:

$$
\begin{array}{c}
\mathrm{H} \\
| \\
\mathrm{H} - \mathrm{C} - \mathrm{H} \\
| \\
\mathrm{H}
\end{array}
$$

Methane (CH_4)

Each dash represents a chemical bond; the carbon atom has four bonds and the hydrogen atom has one bond.

More complex forms of methane can be developed by adhering to the simple rule that a single bond exists between adjacent carbon atoms and that the rest of the bonds are saturated with hydrogen atoms. With the development of more complex forms, molecular size is increased. Hydrocarbons that contain the same number of carbons and hydrogen atoms, but have a different structure, and therefore different properties, are known as isomers. As the number of carbon atoms in the molecule increases, the number of isomers increases rapidly. The simplest hydrocarbon having isomers is butane, which is illustrated as follows:

$$CH_3 - CH_2 - CH_2 - CH_3 \qquad\qquad \begin{array}{c} CH_3 \\ \diagdown \\ \diagup \\ CH_3 \end{array} CH - CH_3$$

Normal Butane   Isobutane

Hydrocarbon compounds can be divided into four major structural forms: (1) alkanes, (2) cycloalkanes, (3) alkenes, and (4) arenes. Petroleum geologists and engineers commonly refer to these structural groups as (1) paraffins, (2) naphthenes or cycloparaffins, (3) olefins, and (4) aromatics, respectively, and they will be referred to as such in this chapter. Olefins are characterized by double bonds between two or more carbon atoms. Because olefins are not found in crude oil and are found only in trace amounts in few petroleum products, since they are readily reduced or polymerized to alkanes early in diagenesis, the following discussions will focus on paraffins, naphthenes or cycloparaffins, and aromatics.

Paraffin-type hydrocarbons are referred to as saturated and aliphatic hydrocarbons. These hydrocarbons dominate gasoline fractions of crude oil and are the principal hydrocarbons in the oldest, most deeply buried reservoirs. Paraffins can form normal (straight) chains and branched-chain structures. Normal chains form a homologous series in which each member differs from the next member by a constant amount, as illustrated in Figure 3.2 in which each hydrocarbon differs from the succeeding member by one carbon and two hydrogen atoms. The naming of normal paraffins is a simple progression using Greek prefixes to identify the total number of carbon atoms present. The common name of some normal-chain paraffins along with their respective physical properties are listed in Table 3.1. Branched-chain paraffins reflect different isomers (different compounds with the same molecular formula). Where only about 60 normal-chain paraffins exist, theoretically, over a million branched-chain structures are possible, with about 600 individual hydrocarbons identified. Common branched-chain paraffins and their respective physical properties are illustrated in Figure 3.2 and are also listed in Table 3.1.

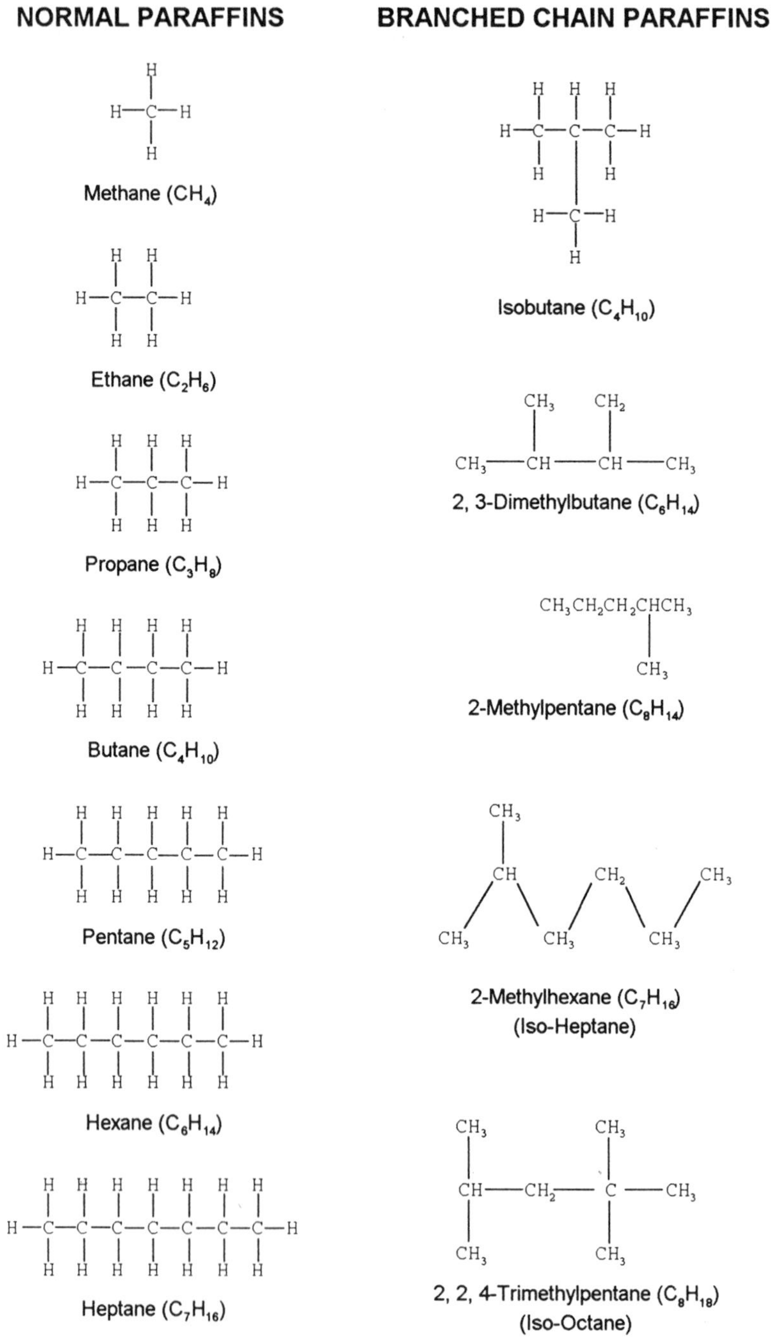

NORMAL PARAFFINS   **BRANCHED CHAIN PARAFFINS**

Methane (CH_4)

Isobutane (C_4H_{10})

Ethane (C_2H_6)

2, 3-Dimethylbutane (C_6H_{14})

Propane (C_3H_8)

$CH_3CH_2CH_2CHCH_3$
CH_3

2-Methylpentane (C_8H_{14})

Butane (C_4H_{10})

Pentane (C_5H_{12})

2-Methylhexane (C_7H_{16})
(Iso-Heptane)

Hexane (C_6H_{14})

Heptane (C_7H_{16})

2, 2, 4-Trimethylpentane (C_8H_{18})
(Iso-Octane)

Figure 3.2.   Structural forms of normal (straight)-chain and branched chain paraffins.

Table 3.1   Chemical and Physical Properties of Common Normal Paraffin and Branched-Chain Paraffins

Compound	Chemical Formula	Molecular Wt.	Density	Solubility ($g/10^6$ g H_2O)	Viscosity (micropoises)	Boiling Point (°C)	Vapor Pressure (mm)
Normal Paraffins							
Methane	CH_4	16	0.554	—	108.7 @ 20°C	−161	400 @ −168.8°C
Ethane	C_2H_6	30	0.446	—	98.7 @ 17°C	−89	400 @ −99.7°C
Propane	C_2H_8	44	0.582	62.4 ± 2.1	79.5 @ 17.9°C	−42	400 @ −55.6°C
Butane	C_4H_{10}	58	0.599	61.4 ± 2.6	—	−0.5	400 @ 16.3°C
Pentane	C_5H_{12}	72	0.626	38.5 ± 2.0	676,000 @ 25°C	36	426
Hexane	C_7H_{16}	86	0.659	9.5 ± 13	3,260 @ 20°C	69	124
Branched-Chain Paraffins							
Isobutane	C_4H_{10}	58	—	48.9 ± 2.1	—	−12	—
2,2-Dimethyl-butane	C_6H_{14}	86	0.649	18.4 ± 1.3	—	50	400
2,3-Dimethyl-butane	C_6H_{14}	86	0.668	22.5 ± 0.4	—	58	400
2-Methylpentane	C_6H_{14}	86	0.669	13.8 ± 0.9	—	60	400 @ 41.6°C
2-Methylhexane	C_7H_{16}	100	0.6789	2.54 ± 0.0	—	90	40
3-Methylhexane	C_7H_{16}	100	—	4.95 ± 0.08	—	92	—
2,2,4-Tri-methylpentane	C_8H_{18}	114	0.692	2.44 ± 0.12	—	99	40.6

Naphthenes or cycloparaffins are formed by joining the carbon atoms in a ring-type structure and are the most common molecular structures in petroleum. These hydrocarbons are also referred to as saturated hydrocarbons, since all the available carbon atoms are saturated with hydrogen. Typical naphthenes and their respective physical properties are listed in Table 3.2 and are shown in Figure 3.3.

Aromatic hydrocarbons usually comprise less than 15% of a total crude oil, although they often exceed 50% in heavier fractions of petroleum. The aromatic fraction of petroleum is the most important environmental group of hydrocarbon chemicals and contains at least one benzene ring comprised of six carbon atoms in which the fourth bond of each carbon atom is shared throughout the ring. Schematically shown with a six-sided ring with an inner circle, the aromatics are unsaturated, allowing them to react to add hydrogen and other elements to the ring. Benzene is known as the parent compound of the aromatic series and, along with toluene, ethylbenzene, and the three isomers of xylene (*ortho-, meta-,* and *para-*) are major constituents of gasoline. Typical aromatics and their respective physical properties are listed in Table 3.3 and shown in Figure 3.4.

The principal method for separating crude oil into useful products is through distillation. Boiling points of hydrocarbons generally increase with an increase in the number of carbon atoms that comprise the compound. As a crude sample (or any hydrocarbon blend) is heated in increasing increments, the hydrocarbon compounds having a boiling point at or below the current temperature volatilizes. The remaining hydrocarbon compounds in the sample will not volatilize until the temperature is raised to their respective boilings. A plot of boiling temperatures (°F) versus cumulative percent volume removed from the sample is referred to as a distillation curve. The range of boiling temperatures are from high to low, divided into the following product types: residue, heavy gas-oil, light gas-oil, kerosene, naphtha, gasoline, and butanes. The crude source has a definite effect on the composition of the refined product. Typical distillation curves for a light gas-oil, kerosene, and a mixture of primarily kerosene and naphtha are shown in Figure 3.5. Major commercial products made by the petroleum industry and associated with distillation are tabulated in Table 3.4. These products correspond to their respective boiling points and carbon ranges as shown in Figure 3.6.

Crude oil and refined petroleum product can easily be differentiated via the interpretation of gas chromatograms using EPA Method 8015. In this analysis, for example, emitted compounds are scanned using a gas chromatograph within the temperature range from 50°C to 300°C at a heating rate of 10°C/min., then held at 300°C for 5 min. The total run time is 30 min. The chromatograms are then compared to standards, calculated, and reported. Standard chromatograms for crude oil, gasoline, naphtha, kerosene, diesel fuel, and JP-5 jet fuel are shown in Figure 3.7.

Table 3.2   Chemical and Physical Properties of Common Napthenes

Compound	Chemical Formula	Molecular Wt.	Density	Solubility (g/10^6 g H_2O)	Viscosity (micropoises)	Boiling Point (°C)	Vapor Pressure (mm)
Methlycyclo-pentane	C_6H_{12}	84	0.749	42 ± 1.6	—	72	—
Cychlohexane	C_6H_{12}	84	0.778	55 ± 2.3	1.02 × 17°C	81	95
Ethylcyclo-phexane	C_8H_{16}	112	—	3.29 ± 0.46	—	132	—
1,1,3-Trimethyl-cyclohexane	C_9H_{18}	126	—	1.77 ± 0.05	—	137	—

NAPTHENES (CYCLOPARAFFINS)

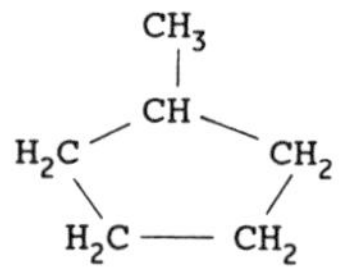

$$CH_3$$
$$CH$$
$$H_2C \qquad CH_2$$
$$H_2C \longrightarrow CH_2$$

Methylcyclopentane (C_6H_{12})

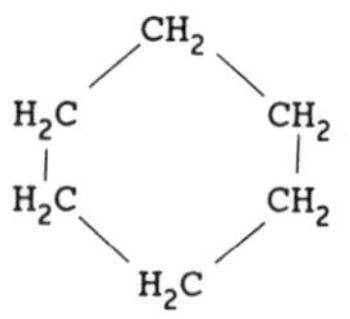

$$CH_2$$
$$H_2C \qquad CH_2$$
$$H_2C \qquad CH_2$$
$$H_2C$$

Cyclohexane (C_6H_{12})

$$CH_2 \qquad CH_2$$
$$H_2C \qquad CH \qquad CH_3$$
$$H_2C \qquad CH_2$$
$$H_2C$$

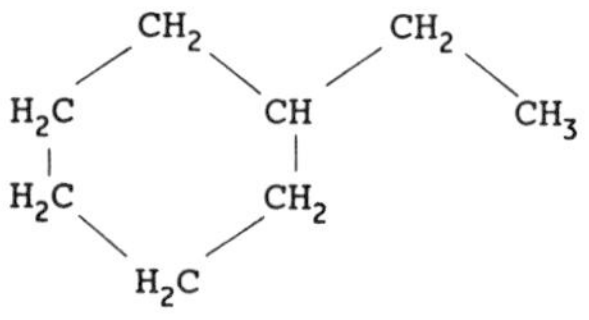

Ethylcyclohexane (C_8H_{16})

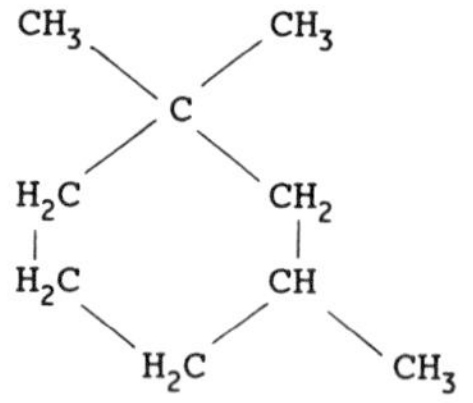

$$CH_3 \qquad CH_3$$
$$C$$
$$H_2C \qquad CH_2$$
$$H_2C \qquad CH$$
$$H_2C \qquad CH_3$$

1, 1,3-Trimethylcyclohexane (C_9H_{18})

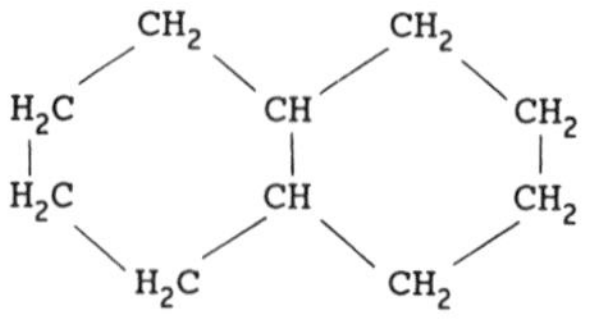

$$CH_2 \qquad CH_2$$
$$H_2C \qquad CH \qquad CH_2$$
$$H_2C \qquad CH \qquad CH_2$$
$$H_2C \qquad CH_2$$

Decalin ($C_{10}H_{18}$)

Figure 3.3.   Structural forms of typical naphthenes.

Table 3.3   Chemical and Physical Properties for Aromatics

Compound	Chemical Formula	Molecular Wt.	Density	Solubility (g/10^6 g H$_2$O)	Viscosity (centipoises) (20°C)	Boiling Point (°C)	Vapor Pressure (mm)	Koc
Benzene	C$_6$H$_6$	78	0.879	1780 ± 45	0.652	80	76 @ 20°C	1.69—2.00
Toluene	C$_7$H$_8$	92	0.866	515 ± 17	0.590	111	22 @ 20°C	2.06—2.18
Ethylbenzene	C$_8$H$_{10}$	106	0.867	161 mg/L at 25°C		136.2°C @ 760 mm/L	4.53 mm at 25°C	3.05—3.15
Othoxylene	C$_8$H$_{10}$	106	0.880	175 ± 8	0.810	142	10 @ 25.9°C	2.11
Metaxylene	C$_8$H$_{10}$	106	0.864	146 ± 1.6	0.620	138.9	10 @ 28.3°C	3.20
Paraxylene	C$_9$H$_{12}$	120	0.861	156 ± 1.6	0.648	138	10 @ 27.3°C	2.31
Isopropyl-benzene	C$_9$H$_{12}$	120	—	50 ± 5	—	152	—	—
3,4-Benzpyrene	C$_{20}$H$_{12}$	252	1.351	0.003	—	495	5.49×10^{-9} @ 25°C	5.81—6.50

AROMATICS

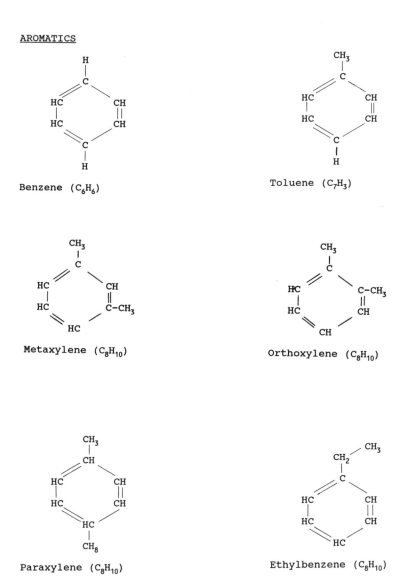

Benzene (C_6H_6)

Toluene (C_7H_3)

Metaxylene (C_8H_{10})

Orthoxylene (C_8H_{10})

Paraxylene (C_8H_{10})

Ethylbenzene (C_8H_{10})

Figure 3.4.   Structural forms of typical aromatics.

3.2   DEGRADATION PROCESSES

Degradation processes include biodegradation as well as chemical and physical degradation via mechanisms such as oxidation, waterwashing, and inspissation (the evaporation of the lighter constituents of petroleum, leaving the

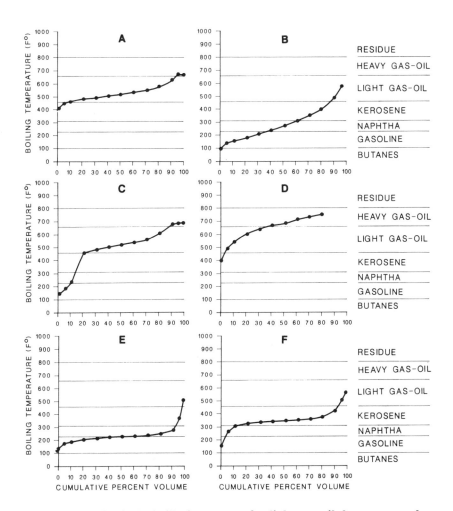

Figure 3.5.   Typical distillation curves for light gas-oil, kerosene, and a mixture of primarily kerosene and naphtha.

heavier residue behind). Degradation processes tend to destroy the paraffins, to remove the light ends, and to oxidize the remaining fractions of the product.

During biodegradation, paraffins in addition to naphthenes and aromatics are all susceptible to microbial decomposition. Over 30 genera and 100 species of various bacteria, fungi, and yeast exist that metabolically utilize one or more kinds of hydrocarbons. The general mechanism for biodegradation is shown in Figure 3.8, in which hydrocarbons are essentially oxidized to alcohols, ketones, and acids.

The order in which hydrocarbons are oxidized depends on numerous factors. However, in general small carbon molecules up to C_{20} are consumed before

Table 3.4   Major Petroleum Distillation Products

Fraction	Distillation Temperature (°C)	Carbon Range
Butanes	—	C_1 to C_4
Gasoline	40–205°	C_5 to C_{14}
Naphtha	60–100°	C_6 to C_7
Kerosene	175–325°	C_9 to C_{18}
Light gas-oil	275°	C_{14} to C_{18}
Heavy gas-oil	275°	C_{19} to C_{25}
Residue	—	$> C_{40}$

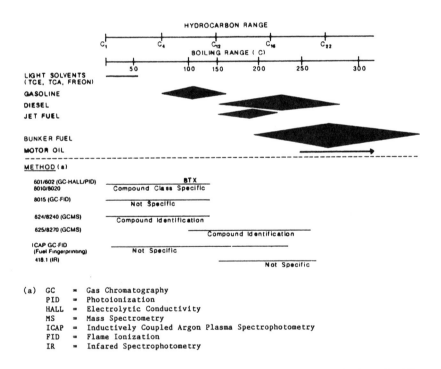

(a) GC = Gas Chromatography
     PID = Photoionization
     HALL = Electrolytic Conductivity
     MS   = Mass Spectrometry
     ICAP = Inductively Coupled Argon Plasma Spectrophotometry
     FID = Flame Ionization
     IR   = Infared Spectrophotometry

Figure 3.6.   Hydrocarbon range, boiling point, and analytical method for major petroleum products.

larger ones. Within the same molecular weight range, the order is usually n-paraffins first. These in turn are followed by isoparaffins, naphthenes, aromatics, and polycyclic aromatics. Because single-ring naphthenes and aromatics are consumed before isoprenoids, steranes, and triterpanes, the remaining diasteranes and tricyclic terpanes survive heavy biodegradation. Thus, they (the diasteranes and tricyclic terpanes) may be useful in chemically fingerprinting biodegraded products.

3.3   GEOCHEMICAL CHARACTERIZATION

Geochemical characterization or "fingerprinting" of leaked crude or refined LNAPL products is important in selecting the appropriate remediation technology. In addition, fingerprinting techniques can also be used as a supportive tool in characterizing type(s) and respective source(s) of hydrocarbons in the subsurface. Fingerprinting techniques in some cases are very conventional in nature, while others were specifically developed for use in the petroleum exploration industry in efforts to characterize source rock and crude oil types. These techniques, in turn, have subsequently been modified and applied to environmental issues, notably in the identification of fugitive hydrocarbons.

Characterization and source identification of subsurface hydrocarbons is made difficult due to several factors, including numerous microbiological, chemical, and physical processes. In addition to these naturally occurring factors are the complex historical development (notably industrial) of a particular site or area, property ownership and transfers over time, and the close proximity of crude and petroleum-handling facilities, ranging from the clustering of refineries, bulk storage tank farms, and underground pipelines in industrialized areas, to numerous underground storage tanks associated with many commercial establishments in less industrialized areas.

Current EPA analytical methods do not allow for the complete speciation of the various hydrocarbon compounds (U.S. EPA, 1986). EPA Methods 418.1 and 8015 provide the "total" amount of petroleum hydrocarbons present. However, only concentrations within a limited hydrocarbon range are applicable to those particular methods. Volatile compounds are usually lost, and samples are typically quantitated against a known hydrocarbon mixture and not the specific hydrocarbon product released. By conducting EPA method 8015 (modified) using a gas chromatograph fitted with a capillary column instead of the standard, hand-packed column, additional separation of various fuel-ranged hydrocarbons can be achieved (California Department of Health Services, 1989).

Several methodologies can be used to identify not only crude or refined product type, but also the brand, grade, and, in some instances, the source crude. The petroleum industry has yielded conventional methods for the characterization of refined products, the simplest being the routine determination of API gravity and the development of distillation curves with LNAPL hydrocarbon. Chemical testing for trace elements, notably lead, manganese, nickel, sulfur, and vanadium, may also be useful in distinguishing product types and crude sources.

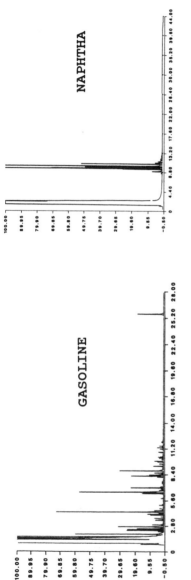

Figure 3.7.   Standard gas chromatograms for crude oil, gasoline, naphtha, kerosene, diesel fuel, and JP-5.

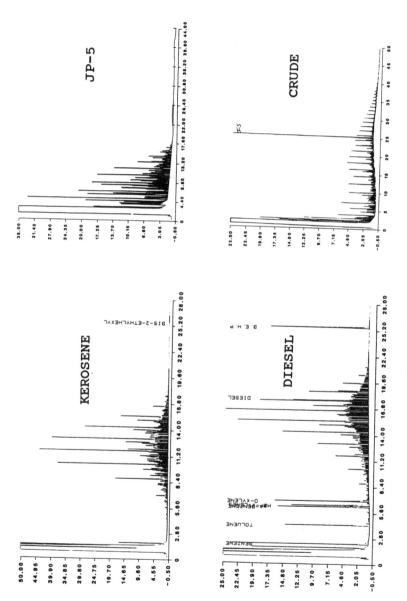

Figure 3.7. continued.

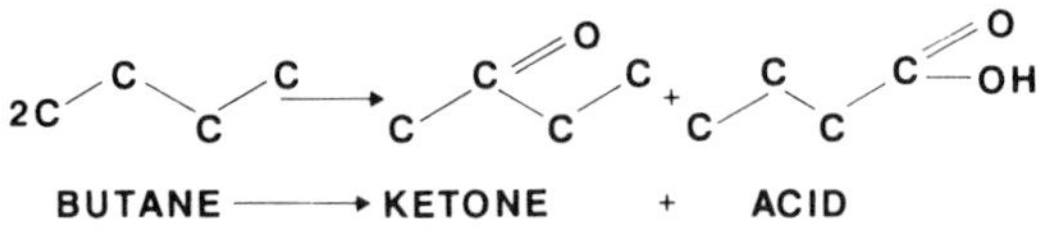

BUTANE ──────→ KETONE   +   ACID

Figure 3.8.   Conversion of *n*-butane to ketone and acid via microbiological oxidation.

More sophisticated methods include gas chromatography, statistical comparisons of the distribution of paraffinic or *n*-alkane compounds between certain C-ranges, and the determination of certain isotopic ratios of carbon and hydrogen for lighter gasoline-range fractions, and ^{15}N and ^{34}S for heavier petroleum fractions. These techniques are discussed below.

3.3.1 API Gravity

Probably the simplest tool in discriminating between LNAPL product types is the measurement of API gravity. API gravity is similar to specific gravity but differs by the following conversion:

$$^\circ API = \frac{141.5}{specific\ gravity} - 131.5 \qquad (3.1)$$

Using this conversion, water has a specific gravity of 1 (unit less) and an API gravity of 10°. The higher the API gravity, the lighter the compound. Typical API gravities for various refined products are listed in Table 3.5. The specific gravity corresponding to API gravity is provided in Appendix A.

The common technique used to measure the API gravity of a liquid sample is by using a hydrometer graduated in degrees of API gravity. The hydrometer is inserted into the liquid sample and allowed to float freely. When equilibrium is reached, the API gravity may be read from the exposed portion of the hydrometer after adjusting for temperature.

Another method to evaluate API gravity commonly applied to oil field evaluations is the use of ultraviolet light having a wavelength of 3600Å. Ultraviolet light of this wavelength causes fluorescence of crude oils in differing colors, which is generally characteristic of the gravity of the oil. Used during drilling, this technique can be used as a quantitative tool for determining API gravity for hydrocarbon-affected soils. Typical colors are presented in Table 3.6.

Plots of boiling temperature (°F) versus cumulative percent volume removed from the sample can be used to distinguish between one or several product types. A graph of typical distillation curves for LNAPL product types retrieved from a refinery site where single but coalescing pools occur is presented in Figure 3.5.

Table 3.5   Typical API Gravities

Product Type	Specific Gravity	API Gravity
Gasoline	0.74	60°
Light crude	0.84	36°
Heavy crude	0.95	18°
Asphalt	0.99	11°

Source: From Leffler (1979).

Table 3.6   Typical Fluoresence Color for Varying API Gravities

API Gravity	Color of Fluorescence
Below 15	Brown
15–25	Orange
25–35	Yellow to cream
35–45	White
Over 45	Blue-white to violet

Source: Exploration Logging, Inc. (1980).

3.3.2 Trace Metals Analysis

The use of tetraethyl and tetramethyl lead as anti-knock compounds in gasoline is widespread, although its use has declined in recent years. Analysis for organic (not total) lead can sometimes serve to identify potential sources of gasoline contamination and timing of releases, if storage history is known. Typically, unleaded gasoline will not contain more than 0.05 g/gal (13 mg/l), while leaded gasoline will not contain greater than 1.1 g/gal (290 mg/l).

During diagenesis of crude oil source materials, both vanadium and nickel become complexed into a porphyrin ring, in addition to other compounds. As porphyrin is introduced into the crude oil from the source rock, it carries the imprint of the original vanadium-nickel distribution. Therefore, concentrations of vanadium and nickel are several thousand times what would be expected if the metals were not complexed into organic structures. Crude oil concentrations of both vanadium and nickel increase an order of magnitude with a decrease in API gravity of about 10°. Vanadium-nickel ratios are often used for oil field cor-relation and may also serve as a supportive tool for environmental investiga-tions. Analysis may be conducted using an atomic absorption spectrophotometer and/or an inductively coupled plasma-mass spectrometer (ICP). The ICP will tend to be more quantitative, reflecting lower achievable detection limits.

3.3.3 Gas Chromatography Fingerprinting

Chromatographic fingerprinting of hydrocarbon samples can be conducted utilizing a gas chromatograph equipped with a capillary column, rather than a standard hand-packed column, to achieve better separation of the compounds as they are eluted, and having a flame ionization detector. Results of hydrocarbons in the range of C_4 to C_8 (gasoline range) may be presented as the relative percent. In the petroleum industry, the ratios of certain compounds or combinations of compounds, are used to indicate the degree of maturity, paraffinicity, biodegradation, and waterwashing.

In a refined product such as gasoline, similar ratios are measurable and may be useful in assessing the similarity of differences in samples, but do not have the same significance as with crude samples due to extensive processing. The values are expected to differ due to the processing (cracking, reforming, etc.) and blending steps the product has undergone. Therefore, gas chromatograph measurements yield a fingerprint that, in principle, can differ from one brand of gasoline to another.

The chromatographic results (fingerprint) of five hydrocarbon samples collected from a monitoring well network located within a refinery tank farm is tabulated in Table 3.7 and is graphically illustrated in Figure 3.9. Results presented are in the range from n-heptane to n-octane. The compositional differences between the five samples are tabulated in Table 3.8. As is readily deduced, samples 1 and 4 are similar to one another, as are samples 2 and 3. This similarity is also evident in Table 3.9.

When a ternary diagram is constructed presenting relative percent "normal" alkanes vs. relative percent branched alkanes (isoalkanes) vs. relative percent cyclic alkanes, the similarity is even more striking (Figure 3.10). The grouping of sample number 1 with 4 and sample number 2 with 3 is strong, while the dissimilarity of sample number 5 from the other four samples is also clear. Additional aromatic compound distribution can be discerned using gas chromatographs, as illustrated in Figure 3.11.

3.3.4 Isotope Fingerprinting

Ratios of the stable isotopes of carbon ($^{13}C/^{12}C$) and hydrogen ($^2H/^1H$) differ both among crude oils by origin and among various fractions within the same oil. The great utility of isotope ratios is their application as a tracer for organic

Table 3.7   Gasoline Range Condensate Hydrocarbon Analysis in
Relative Percent

Name	Sample Number				
	1	2	3	4	5
11 2.3-Dimethylbutane	1.28	0.55	0.17	1.28	3.80
12 2-Methylpentane	4.37	1.54	1.02	5.43	18.23
13 3-Methylpentane	3.38	1.49	1.17	4.54	10.93
14 n-Hexane	4.36	1.28	1.16	5.65	6.48
15 2,2-Dimethylpentane					
16 Methylcyclopentane	9.66	5.53	5.04	10.01	10.15
17 1,4-Dimethylpentane	1.22	0.23	0.19	0.74	0.34
18 2,2,3-Trimethylbutane					
19 Benzene		2.87	3.25		3.53
20 3,3-Dimethylpentane		0.04	0.34		
21 Cyclohexane	4.72	3.90	5.01	4.21	2.48
22 2-Methylhexane	5.39	1.83	1.14	4.33	1.73
23 2,3-Dimethylpentane	1.51	1.24	1.16	1.73	0.61
24 1,1-Dimethylcyclopentane		0.93	0.96		0.36
25 3-Methylhexane	3.60	3.09	2.37	3.22	1.48
26 c-1,3-Dimethylcyclopentane	4.14	3.25	2.92	4.24	1.40
27 t-1,3-Dimethylcyclopentane	3.96	3.18	2.90	3.90	1.46
28 t-1,2-Dimethylcyclopentane	8.17	6.34	5.74	8.07	2.50
29 n-Heptane	4.37	3.51	2.56	4.73	3.18
30 Methylcyclohexane	10.91	13.04	12.96	11.6	5.07
31 Trimethylcyclopentanet(iso?)	1.72	1.94	1.88	1.78	0.79
32 Ethylcyclopentane	2.02	2.10	1.86	1.77	1.02
33 2,5-Dimethylhexane	0.93	0.61	0.46	.062	
34 2,4-Dimethylhexane	0.96	0.82	0.72	0.89	
35 t-1,2-c-5-Trimethylclopentane	1.90	2.79	2.76	2.09	1.63
36 1,1,2-Trimethylcyclopentane	3.48	5.14	4.65	4.16	1.97
37 2,3,4-Trimethylpentane	1.38	0.00		1.18	
38 Toluene	1.52	0.00		1.13	
39 2-Methylheptane	5.61	10.59	15.25	5.46	6.56
42 Dimethlcyclohexane(iso?)					
43 Dimethlcyclohexane(iso?)	1.10	2.22	1.88	1.42	
44 n-Octane	4.03	11.46	12.60	1.77	9.54

Blanks in this table mean no peak was detected. Data that are 0.00 were detected
at levels less than 0.005%.

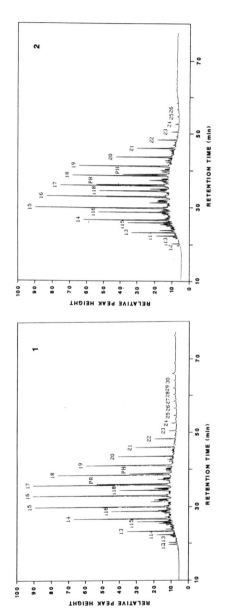

Figure 3.9.   Gas-chromatograph results of crude and refined LNAPL hydrocarbon product samples retrieved from a monitoring well network located within a refinery tank farm.

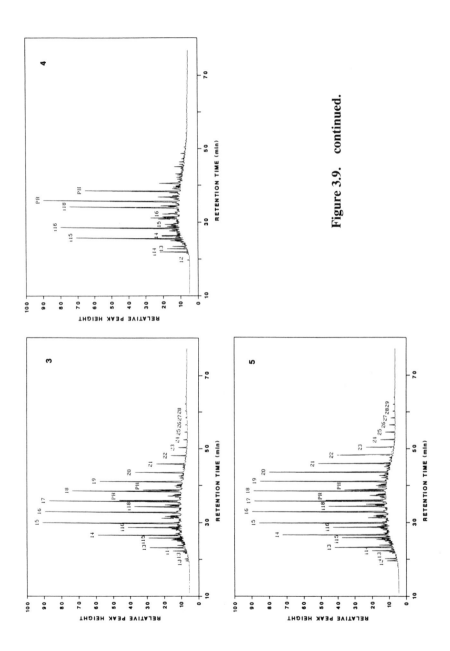

Figure 3.9.   continued.

Table 3.8   Gasoline Range Condensate Hydrocarbon Ratios

		Sample Number				
Name		**1**	**2**	**3**	**4**	**5**
A 19/30	Late maturity Index	0.00	0.22	0.25	0.00	0.70
B 38/29	Aromaticity	0.35	0.00	0.00	0.24	0.00
C (1)	Paraffinicity	0.56	0.28	0.21	0.66	1.28
F 29/30	Paraffinicity	0.40	0.27	0.20	0.41	0.63
I (2)	Isoheptane value	0.55	0.39	0.30	0.47	0.60
H (3)	Heptane value	9.66	9.00	6.99	10.68	16.17
K (5)	Kerogen type index	10.24	12.96	9.87	14.36	20.34
U 21/16	Napthene branching	0.49	0.71	1.00	0.42	0.24
R 29/22	Paraffin branching	0.81	1.93	2.24	1.09	1.84
13/19	Waterwashing parameter	0.52	0.36			3.10
30/38	Waterwashing parameter	7.16			10.33	
13/14	Biodegradation parameter	0.77	1.16	1.00	0.80	1.69
	Rel % Normal alkanes	12.95	16.74	16.87	12.29	19.90
	Rel % Branched alkanes	34.49	31.42	32.95	33.85	50.23
	Rel % Cyclic alkanes	52.55	51.84	50.18	53.86	29.87
	Total	99.99	100.00	100.00	100.00	100.00

(1) (14+29)/(21+30)
(2) (22+25)/(26+27+28)
(3) (100.0*29/(21+22+24+25+26+27+28+29+30))

materials that have undergone partial degradation through either biological or other processes in which the molecular characteristics of the original substance are obscure. This is because the isotopic composition does not change to the same extent as the molecular composition.

The actual ratio of ^{13}C to ^{12}C is determined on an isotope ratio mass spectrometer; however, this actual ratio is then compared to a standard using the following equation (Kaplan, 1989) to calculate the ratio difference in parts per thousand:

$$\delta^{13}C^{\circ}/^{\circ\circ} = \frac{\left(^{13}C/^{12}C \text{ sample} - \,^{13}C/^{12}C \text{ standard}\right)}{^{13}C/^{12}C \text{ standard}} \times 1000 \qquad (3.2)$$

The most widely used standard is a belemnite from the Peedee formation in South Carolina (PDB); therefore, some ratios may be expressed as negative values.

Table 3.9   Comparison of Composition Differences

Name	Difference in Hydrocarbon Ratios[a]
A 19/30 Late maturity index	–0.70
B 38/29 Aromaticity	0.24
C (1) Paraffinicity	–0.62
F 29/30 Paraffinicity	–0.22
I (2) Isoheptane value	–0.13
H (3) Hepetane value	–5.48
K (4) Kerogen type index	–5.99
U 21/16 Naphthene branching	0.18
R 29/22 Paraffin branching	–0.74
13/19 Waterwashing parameter	–3.10
30/38 Waterwashing parameter	10.33
13/14 Biodegradation parameter	–0.89
Rel % normal alkanes	–7.61
Rel % branched alkanes	–16.38
Rel % cyclic alkanes	23.99
Number of ratio differences < 1.0	8
Number of ratio differences < 4.0	3

[a] Calculated by subtracting hydrocarbon ratio of sample 5 from sample 4.

Most carbon isotope ratio correlations are made on the ^{15+}C fraction of crude oil, because it is less affected by degradation processes. Valid correlations using carbon isotopes can only be conducted on the same fractions of samples.

Carbon and hydrogen isotopes are measured as a ratio of the heavier to the lighter (most abundant) isotope in a gas introduced into a dual-collecting mass spectrometer. The separated and purified petroleum product is combusted under a vacuum with an oxidizing catalyst to produce CO_2 and H_2O. The CO_2 is then purified and injected into the mass spectrometer to measure the masses $44(^{12}C^{16}O_2)$ and $45(^{13}C^{16}O_2)$. The $^{13}C/^{12}C$ is then measured relative to the PDB standard. The water produced during the combustion of the hydrocarbon sample is reacted with zinc metal and converted to H_2 which is then introduced to a mass spectrometer where masses $2(^{2}H)$ and $3(DH)$ are measured. The deuterium/hydrogen ratio thus obtained is compared relative to an international standard.

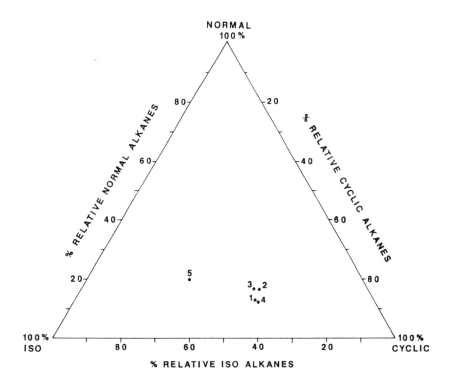

Figure 3.10.   Ternary diagram showing relative percentages of normal alkanes, branched-alkanes (iso-alkanes), and cyclic alkanes based on gas chromatograph results on LNAPL hydrocarbon product samples.

A two-component plot of the $^{13}C/^{12}C$ isotope ratio from the saturated and aromatic fractions of the five samples previously identified is presented in Figure 3.12. The data indicates that samples 2, 4, and 5 are clearly related to one another, as are samples 1 and 3, indicating similar source crude oils.

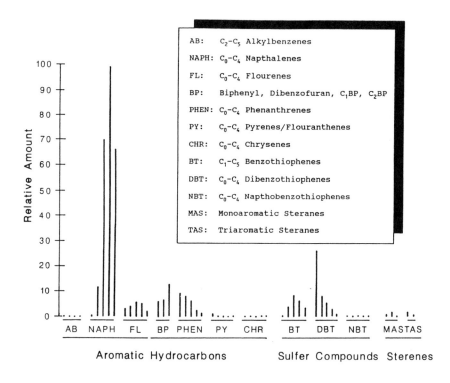

Figure 3.11. Additional aromatic compound, sulfur and sterenes distribution based on gas chromatograph results.

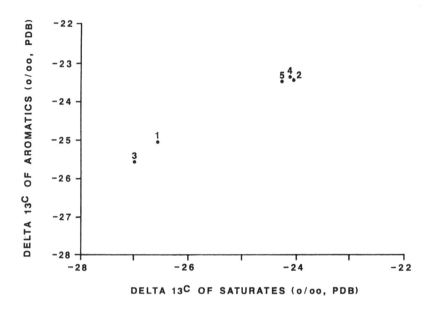

Figure 3.12   Two-component plot of the $^{13}C/^{12}C$ isotope ratio from the saturated and aromatic fractions in five LNAPL hydrocarbon product samples.

REFERENCES

1. Baker, E. W., 1964, Vanadium and nickel in crude petroleum of South America and Middle East Origin: *J. Chem. Eng. News*, v. 42, (15), p. 307-308.
2. Bland, W. F. and Davidson, R. L., 1967, *Petroleum Processing Handbook:* McGraw-Hill Book Company, New York, New York.
3. Bowie, C. P., 1918, Oil-Storage Tanks and Reservoirs with a Brief Discussion of Losses of Oil in Storage and Methods of Prevention: *U.S. Dep. Int. Bur. Mines Bull.*, No. 155, Petroleum Technology No. 41, 76 p.
4. Brookman, G. T., Flanagan, M., and Kebe, J. O., 1985, *Literature Survey: Unassisted Natural Mechanisms to Reduce Concentrations of Soluble Gasoline Components:* American Petroleum Institute of Health and Environmental Sciences Department, No. 4415, August, 1985, 73 p.
5. Brookman, G. T., Flanagan, M., and Kebe, J. O., 1985, *Literature Survey: Hydrocarbon Solubilities and Attenuation Mechanisms:* American Petroleum Institute of Health and Environmental Sciences Department, No. 4414, August, 1985, 101 p.

6. California Department of Health Services, 1989, Total Petroleum Hydro-carbons (TPH) Analysis - Gasoline and Diesel: in *California Water Re-sources Control Board Leaking Underground Fuel Tank (LUFT) Manual,* Appendix C.
7. Dragun, J., 1988, *The Soil Chemistry of Hazardous Materials:* Hazardous Materials Control Research Institute, Silver Springs, Maryland, 458 p.
8. Exploration Logging, Inc., 1980, *Field Geologists Training Guide, An Introduction to Oilfield Geology, Mud Logging and Formation Evalu-ation,* p. 5-46 to 5-47.
9. Guthrie, V. B., 1960, *The Petroleum Products Handbook:* McGraw-Hill Publishing, New York, New York.
10. Hawley, G., 1981, *The Condensed Chemical Dictionary,* 10th Ed.: Van Nostrand Reinhold Company, Inc., New York, New York.
11. Hodgson, G. W., 1954, Vanadium, Nickel and Iron Trace Metals in Crude Oils of Western Canada: *Am. Ass. Pet. Geol. Bull.,* 38, p. 2537-2554.
12. Howard, P. H., 1989, *Handbook of Environmental Fate and Exposure Data for Organic Chemicals, Volume I,* Large Production and Priority Pol-lutants: Lewis Publishers, Chelsea, MI, 574 p.
13. Hunt, J. M., 1979, *Petroleum Geochemistry and Geology:* W. H. Freeman and Company, San Francisco, California, p. 617.
14. Kaplan, I. R., 1989, Forensic Geochemistry in Characterization of Petro-leum Contaminants in Soils and Groundwater: in *Environmental Con-cerns in the Petroleum Industry* (Edited by Testa, S.M.), Pacific Section of the American Association of Petroleum Geologists, Symposium Vol-ume, p. 159-181.
15. Kraemer, A. J. and Calkin, L. P., 1925, Properties of Typical Crude Oils from the Producing Fields of the Western Hemisphere: *U.S. Bur. Mines Tech. Paper,* 346, p. 1-43.
16. Kraemer, A. J. and Lane, E. C., 1937, Properties of Typical Crude Oils from Fields of the Eastern Hemisphere: *U. S. Bur. Mines Tech. Paper,* 401, p. 1-169.
17. Leffler, W. L., 1979, *Petroleum Refining for the Non-Technical Person:* PennWell Books, Tulsa, Oklahoma, 159 p.
18. McAuliffe, C., 1966, Solubility in Water of Paraffin, Cycloparaffin, Oleofin, Acetylene, Cycloolefin and Aromatic Hydrocarbons: *Jour. Phys. Chem.,* Vol. 70, p. 1267-1275.
19. Mair, B. J., 1967, *Annual Report for the Year Ending June 30, 1967:* American Petroleum Institute Research Project 6, Pittsburgh, Pennsylva-nia, Carnegie Institute of Technology.
20. Montgomery, J. H. and Welkom, L. M., 1990, *Groundwater Chemical Desk Reference:* Lewis Publishers, Chelsea, MI, 640 p.
21. Morrison, R. T. and Boyd, R. N., 1973, *Organic Chemistry:* 3rd Ed., Van Nostrand Reinhold Company, New York, New York, 1258 p.

22. Nelson, W. L., 1941, *Petroleum Refinery Engineering:* McGraw-Hill Book Company, 2nd Ed., New York, New York, 215 p.

23. Nyer, E. K. and Skladany, G. J., 1989, Relating the Physical and Chemical Properties of Petroleum Hydrocarbons to Soil and Aquifer Remediation: *Ground Water Monitoring Rev.,* Winter Issue, p. 54-60.

24. Perry, J. T., 1984, Microbial Metabolism of Cyclic Alkanes: in *Petroleum Microbiology* (Edited by Atlas, R.M.), MacMillan Publishing Co., New York, p. 61-98.

25. Petrov, A. A., 1987, *Petroleum Hydrocarbons:* Springer-Verlag, New York, New York, 255 p.

26. Rosscup, R. J. and Bowman, J., 1967, *Thermal Stabilities of Vanadium and Petroporphyrins:* Preprints of the Division of Petroleum Chemistry, American Chemical Society, Vol. 12, 77 p.

27. Rossini, F. D., 1960, Hydrocarbons in Petroleum: *J. Chem. Ed.,* Vol. 37(11), p. 554-561.

28. Seifirt, W. K. and Moldowan, J. M., 1979, Applications of Steranes, Terpanes and Monoaromatics to the Maturation, Migration and Source of Crude Oils: *Geochim. Cosmochim. Acta,* Vol. 42(1), p. 77-95.

29. Seifirt, W. K. and Moldowan, J. M., 1979, The Effects of Biodegradation on Steranes and Terpanes in Crude Oils: *Geochim. Cosmochim. Acta,* Vol. 43(1), p. 111-126

30. Testa, S. M., Henry, E. C., and Hayes, D., 1988, Impact of the Newport-Inglewood Structural Zone on Hydrogeologic Mitigation Efforts - Los Angeles Basin, California: in *Proceedings of the National Water Well Association of Ground Water Scientists and Engineers FOCUS Conference on Southwestern Groundwater Issues,* p. 181-203.

31. Testa, S. M., Baker, D. M., Avery, P. L., 1989, Field Studies on Occurrence, Recoverability and Mitigation Strategy for Free Phase Liquid Hydrocarbon: in *Environmental Concerns in the Petroleum Industry* (Edited by Testa, S.M.), Pacific Section of the American Association of Petroleum Geologists Symposium Volume, p. 57-81.

32. Testa, S. M., 1990, Hydrocarbon Product Characterization: Applications and Techniques: in *Proceedings of the National Water Well Association of Groundwater Scientiests and Engineers Fourth Outdoor Action Conference on Aquifer Restoration, Ground Water Monitoring and Geophysical Methods,* May, 1990, in press.

33. Testa, S. M. and Halbert, W. E., 1989, Geochemical Fingerprinting of Free Phase Liquid Hydrocarbons: in *Proceedings of the National Water Well Association and American Petroleum Institute Conference on Petroleum Hydrocarbons and Organic Chemicals in Ground Water: Prevention, Detection and Restoration,* NWWA, Houston, Texas, p. 29-44.

34. United States Environmental Protection Agency, 1986, *Test Methods for Evaluating Solid Waste; Physical/Chemical Methods.* SW-846, 3rd Ed. Office of Solid Waste and Emergency Response, U.S. EPA, Washington, D.C.

4 FATE AND TRANSPORT OF SPILLED/LEAKED PETROLEUM IN THE SUBSURFACE

"Methodology must be developed to predict where in the environment a chemical will be transported, the rate and extent of transformation, and its effect on organisms and environmental processes at expected ambient levels"

4.1 INTRODUCTION

When a significant quantity of liquid petroleum hydrocarbon is released into the subsurface (i.e., underground storage tank, piping, etc.), several migration pathways exist. Migration of liquid petroleum into the subsurface can be divided into three stages as follows:

• Seepage through the unsaturated zone
• Spreading over the water table (referred to as the pancake layer)
• Stability within the water capillary zone

Once a significant volume of liquid hydrocarbons is released, the hydrocarbon migrates downward generally under the influence of gravity and subordinate capillary forces until it reaches the capillary fringe above the water table. In homogeneous, isotropic materials, seepage occurs with minimal amounts of lateral spreading as shown in Figure 4.1a. Primary factors affecting the amount of lateral spreading include the rate of release, the volume of release, and the presence of significant permeability contrasts, as would be anticipated in heterogeneous anisotropic materials. For example, a large instantaneous release

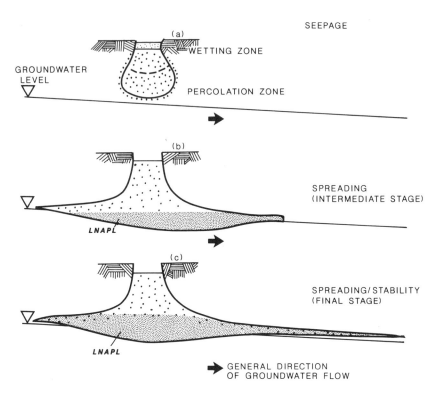

Figure 4.1.   Stages of subsurface hydrocarbon migration (after Schwille, 1967).

into the unsaturated zone will have a higher degree of spreading in comparison to a continuous small release. More prevalent slow releases typical of underground storage tanks and associated piping may also develop fingers of product migrating vertically downward, as shown in Figure 4.2 (Van Duijvenbooden and Kooper, 1981). Where soil or rock heterogeneities and anisotropy exists, as is most often the case, flow paths can be substantially different from those depicted. This phenomenon is common when zones of significant permeability contrasts are encountered (i.e., clay perched zones), as illustrated in Figure 4.2. To complicate matters, lateral seepage within the unsaturated zone may in some cases occur in directions contrary to the overall direction of groundwater flow due to geologic conditions such as stratification, channeling, bedrock orientation, etc.

Residual saturation capacity of soil is generally about one third that of their water-holding capacity (Bossert et al., 1984). Immobilization of a certain mass of hydrocarbon is dependent upon soil porosity and physical characteristics of the product. The volume of soil required to immobilize a volume of liquid hydrocarbon can be estimated (American Petroleum Institute, 1972) as follows:

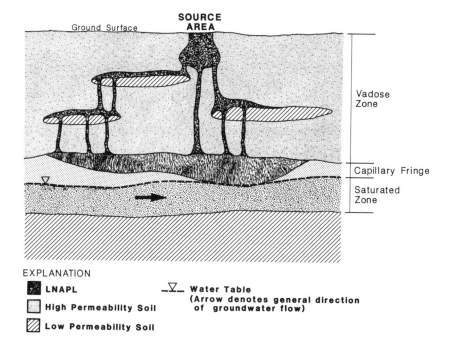

EXPLANATION

■ LNAPL

☐ High Permeability Soil

▨ Low Permeability Soil

▽ Water Table
(Arrow denotes general direction
of groundwater flow)

Figure 4.2. Schematic showing distribution of LNAPL in the subsurface.

$$V_s = \frac{0.2\ V_{hc}}{P(RS)},$$
(4.1)

where V_s = cubic yards of soil required to attain residual saturation; V_{hc} = volume of discharged hydrocarbon, in barrels (42 gal = 1 barrel); P = soil porosity; and RS = residual saturation capacity. The maximum depth of liquid hydrocarbon penetration into the unsaturated zone can be estimated. A nonrigorous approach is presented by Dietz (1970) and Van Dam (1967) as follows:

$$D = \frac{V_s}{A},$$
(4.2)

where D = maximum depth of liquid hydrocarbon penetration into the unsaturated zone; V_s = cubic yards of soil required to attain residual saturation; and A = area of infiltration. CONCAWE (1974) presents another alternative nonrigorous approach whereas:

$$D = \frac{KV_{hc}}{A},$$
(4.3)

Table 4.1   Typical K Values for Soils

Soil Type	K Value		
	Gasoline	Kerosene	Light Gas Oil
Stone to coarse gravel	400	200	100
Gravel to coarse sand	250	125	62
Coarse to medium sand	130	66	33
Medium to fine sand	80	40	20
Fine sand silt	50	25	18

Source: Dietz (1970) and CONCAWE (1974).

Table 4.2   Typical Oil Retention Capacities for Kerosene in Unsaturated Soils

Soil Type	Oil Retention Capacity (R)	
	l/m³	g/yd³
Stone, coarse sand	5	1
Gravel, coarse sand	8	2
Coarse sand, medium sand	15	3
Medium sand, fine sand	25	5
Fine sand, silt	40	8

Source: CONCAWE (1979).

where D = maximum penetration depth of liquid hydrocarbon in the unsaturated zone; K = constant based on soil retention capacity for oil and oil viscosity of oil (Table 4.1); V_{hc} = volume of discharged hydrocarbon in barrels (42 gal = 1 barrel); A = area of infiltration. A third nonrigorous approach is also presented by CONCAWE (1979) as follows:

$$D = \frac{1000\ V_{hc}}{ARC},$$ (4.4)

where D = maximum penetration depth of liquid hydrocarbon in the unsaturated zone; V_{hc} = volume of discharged hydrocarbon in barrels (42 gal = 1 barrel); A = area of infiltration; R = soil retention capacity (Table 4.2); and C = approximate correction factor based on product viscosity (0.5 for gasoline to 2.0 for light fuel oil).

When vertically migrating LNAPL hydrocarbon nears the water table, the capillary fringe is initially encountered. This capillary zone rises above the water table to a height dependent upon the grain size distribution the formation as discussed later in this chapter. Essentially, finer grained soils such as silt or clay attain a thicker capillary zone than coarser grained soil, such as sand or gravel. As the light hydrocarbon enters the water capillary zone, it begins to fill the pore spaces not occupied by capillary or residual water. Little mixing occurs since the two fluids are immiscible. Upon reaching the top of the water capillary zone, additional light hydrocarbon accumulation begins to spread laterally to form what is referred to as a pancake (Figure 4.1b). The lateral spreading of the pancake layer over the water table can be estimated (CONCAWE, 1974) as follows:

$$S = \left(\frac{1000}{F} \right) \left(V - \left[\frac{Ad}{K} \right] \right), \qquad (4.5)$$

where S = maximum spread of the pancake (m^2); F = thickness of the pancake (mm); V = volume of infiltrating bulk hydrocarbons (m^3); A = area of infiltration (m^2); d = depth to groundwater (m); K = constant dependent upon the soil retention capacity for oil and upon oil viscosity (Table 4.1).

K values generally decrease with decreasing grain size. Typical values for K are provided by Dietz (1970) and CONCAWE (1974) and are summarized in Table 4.1.

Sufficiently large seepage rates typically produce a hydraulic mound which permits limited lateral spreading of hydrocarbon in directions inconsistent with the original water table gradient, although lateral spreading upgradient could also be stratigraphically controlled. The initial stage of lateral spreading is dominated by gravity forces, but as the gravitational potential diminishes, capillary forces tend to control the rate of lateral spread.

Where the source is limited or ceases, capillary spreading eventually slows until further migration is limited and equilibrium is reached (Figure 4.1c). This stable condition is attained when the leading edge of the laterally spreading light LNAPL fails to be replenished by more hydrocarbon. In this stable condition, the formation reaches its immobile or residual hydrocarbon saturation.

Residual hydrocarbon saturation will exist within both the unsaturated and capillary zone through which the LNAPL phase migrated. As might be expected, residual hydrocarbon saturation tends to be higher as the grain size decreases and the hydrocarbon viscosity increases. For example, hydrocarbons become immobile at a much higher hydrocarbon saturation in clay than in sand. As a practical matter, this means that following hydrocarbon recovery efforts, more hydrocarbon is retained in clayey formations than in sandy formations.

The basic principles governing downward migration of light hydrocarbons as discussed above are applicable to "dense" hydrocarbons (DNAPL) as well. The difference is that once groundwater is encountered, dense hydrocarbons con-

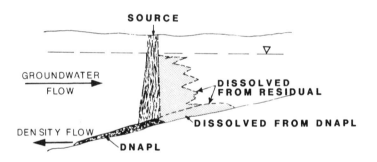

Figure 4.3.   Schematic showing distribution of DNAPL in the subsurface.

tinue to migrate downward, reflecting a specific gravity or density greater than that of water. Although a pancake may initially form, downward migration occurs once sufficient mass is attained. The migration of dense hydrocarbon in the subsurface is schematically shown in Figure 4.3.

The phenomenon of flow of DNAPL hydrocarbons in the saturated zone is referred to as density flow. Density flow primarily occurs by front displacement, whereas the leading edge of the plume displaces groundwater. Fingering can also occur in the saturated zone, reflecting density and viscosity differences (Duijvenbooden and Kooper, 1981). Vertical density flow will cease when a significant permeability contrast is encountered (i.e., clay) or when the density flow is transformed into residual hydrocarbon saturation.

This simplistic mental image depicting the flow path of leaked liquid introduces the concepts of what processes are involved, how does the flow occur, and what is the ultimate fate of the hydrocarbon liquid. Major processes operating to assimilate spilled petroleum liquids in the subsurface, both above and below the water table, are also discussed. These processes include volatilization, adsorption, immiscible fluid flow, dispersion (in solution), and biodegradation.

4.2   VOLATILIZATION

The presence of petroleum vapors in the shallow subsurface is an indication that there is (or was) product that has volatilized. The source of vapors can be a free-phase pool, product that is adsorbed onto soil particles, or product dissolved in the groundwater. After the product has volatilized into the air phase, it moves by dispersion from areas of higher to lower concentration. Ultimately it may reach the surface. If the air in the soil is moving, the rate of transfer can be much greater (by the process of convection). However, the movement of air in the unsaturated zone is very slow under natural conditions, resulting in a long time of contact between the phases. Often, a concentration equilibrium is reached near the phase boundary.

After vapor is dispersed through air in the soil pores, its fate may include escape to the atmosphere, adsorption by soil particles, destruction by biodegradation, or re-solution into percolating rainwater. In highly permeable unsaturated settings, volatilization of light hydrocarbons (i.e., gasoline) can be a major factor in soil remediation. Similarly, where large quantities of gasoline are located on a shallow water table aquifer under a building (or adjacent to underground structures), the potential for explosion is a real danger.

Several soil vapor monitoring techniques are currently being used to define areas of petroleum contamination. These procedures usually involve the collection of representative samples of the soil gas for analysis of indicator compounds. Maps marked with concentration contours of these indicator compounds can be used to identify potential sources to delineate the contaminated area. Indicator compounds are selected for each specific situation, and usually the more volatile components are used. For gasoline contamination, the compounds are usually benzene, toluene, ethylbenzene, and total xylene. In the case of a fuel oil spill, the most commonly used indicator is naphthalene.

Utilization of this mapping technique can be an effective investigative tool when it is used in conjunction with a detailed subsurface investigation. The presence of confining layers, permeable pathways, or previous spills may lead the investigator into erroneous conclusions. Accurate interpretation of soil gas survey data is a specialized discipline in the science of remediation.

The volatility of some fuel products can be a great aid in remediation. Under favorable geological conditions, a system of vapor extraction wells can be a highly effective recovery technique, as discussed later in Chapter 9.

4.3   ADSORPTION

The term *adsorption* describes the process by which molecules (or ions) contained in a liquid or gaseous phase tend to concentrate on a solid phase surface. In soils, this process is exhibited when molecules of gas, free liquid product, or contaminants dissolved in water are attached to the *surface* of an individual soil particle (often in the form of organic carbon). This surface attachment can be any of three general types:

- *Physical*: very weak, caused by Van der Waals forces
- *Chemical*: much stronger, similar in strength to some chemical bonds; often requires significant effort to separate
- *Exchange adsorption*: characterized by electrical attraction between the *adsorbate* and the surface. This type of adsorption is exemplified by ion exchange processes

Adsorptive reactions of LNAPL and dissolved organic compounds moving through soil are almost always of the chemical type, which are reversible equilibrium reactions. A concentration balance is reached between that existing in the liquid (or dissolved) phase and that which is attached to the soil particles. When concentration conditions change, the soil may adsorb additional organic molecules or release them.

Since adsorption is a surface phenomenon, its activity is a direct function of the surface area of the solid as well as the electrical forces active on that surface. Most petroleum chemicals are nonionic and therefore associate more readily with organic than with mineral particles in soils. Dispersed organic carbon found in soils has a very high surface to volume ratio. A small percentage of organic carbon can have a larger adsorptive capacity than the total of the mineral components.

Adsorption is especially important in sediments containing a high percentage of organic matter (peat bogs, former lake beds, etc.). Mass migration of product through these soils is greatly retarded. Most aquifers, however, contain less than 0.1% carbon content and the rate of adsorption is minimal.

Adsorption isotherms for relatively low concentrations of hydrophobic compounds (i.e., less than one half solubility) are linear and may be described by the following equation:

$$q = k_d \, C, \tag{4.6}$$

where q = mass adsorbed per mass of solid phase; k_d = slope of the adsorption isotherm; and C = equilibrium concentration of the contaminant dissolved in water. As this equation indicates, the adsorption of a contaminant from water onto carbon is dependent upon the molecule being adsorbed. Less soluble molecules are more easily adsorbed. Another factor that affects the slope of the isotherm (for any given pressure and temperature) is the molecule and its ability to fit within the pore space of the carbon.

Definition of K_d is difficult to establish when dealing with real field situations. Laboratory studies have demonstrated that an approximation of K_d can be made based on the more easily determined factors such as organic carbon content of the soil and the octanol-water partitioning coefficient of the compound. For sediment particles <50 μm (microns):

$$K_d = .6 \, f_{oc} \, K_{ow},\qquad(4.7)$$

where K_d = slope of the adsorption isotherm; f_{oc} = fractional organic carbon in the solid phase; and K_{ow} = octanol water partition coefficient. K_{ow} for many compounds has been determined by measuring the ratio of the concentration of a particular compound in octanol (a relatively nonpolar alcohol) to that in water. This ratio is also often used to approximate the partitioning behavior of compounds between soil-containing organic matter and water.

The net result of adsorption in petroleum contaminated soils and aquifers is to retard the movement of contaminants. When a pollutant is adsorbed onto soil, it can be released only when the equilibrium between it and the passing water (or air) is disrupted. If the seepage velocity of uncontaminated water passing through the system is too fast, the water will contain less than the equilibrium concentration of the contaminant. Monitoring wells in dynamic flow settings may demonstrate low concentrations of dissolved contaminants during periods of rapid flow. When slower flow is again present, concentrations will increase to the equilibrium level.

4.4   OCCURRENCE AND FLOW OF IMMISCIBLE LIQUIDS

4.4.1   The Unsaturated Zone

4.4.1.1   Static Conditions

Water (or other liquids) present in and moving through the unsaturated zone is subjected to several additional rules of physics than water below the water table. The presence of retained moisture above the water table is due to adsorptive forces between the water molecules and soil particles, in addition to surface tension of the water surface. The term *capillary action* describes the upward movement of a fluid due to surface tension through pore spaces. The fluid can rise until the lifting forces are balanced by gravitational pull (Figure 4.4).

The rise of fluid in a small tube above an LNAPL surface can be described by the equation:

$$h = \frac{2T\cos\theta}{rpg},\qquad(4.8)$$

where h = rise in cm; T = surface tension in dynes; θ = wall-liquid interface angle; p = density of fluid; g = acceleration of gravity; and r = radius of tube in centimeters.

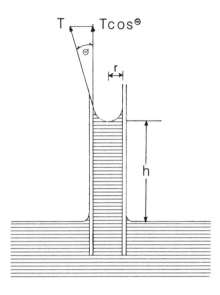

Figure 4.4.   Rise of liquid in a capillary tube.

Lifting of fluids above the LNAPL surface is a true negative pressure compared to atmospheric pressure (also described as soil suction). In soil situations, pore spaces are not uniform elongated cylinders and, therefore, the actual rise of fluid does not create a smooth uniform front. Capillary forces can also promote downward or horizontal fluid migration.

The occurrence of these principles in soil above the water table is depicted in Figure 4.5. Immediately above the water table, the great majority of the pore spaces are filled with water. This condition extends upward to form the capillary fringe which is virtually saturated. Next in upward progression is the funicular zone, in which the volume of water retained is reduced as the number of directly connected small pores decreases and the percentage of air saturation increases. Above the funicular zone is the pendular zone, in which water is retained as "residual saturation" in the necks of individual pores. This situation is very stable because adhesive forces per volume retaining the remaining water are much greater than the gravity draining force. Under some circumstances the adhesive forces are as great as several atmospheres. Water in the liquid state cannot exist under a suction of greater than 0.7 atm.

Although water is the most common liquid fluid found in the unsaturated zone, other liquids that may occur there are controlled by the same physical forces. Laboratory testing of diesel fuel and water in sands demonstrate that free-floating oil above a water also creates a capillary fringe. Because the surface tension of oil is characteristically less than water, the oil-capillary fringe for the diesel fuel used in the laboratory experiments was approximately one half as high as that of the water-capillary fringe.

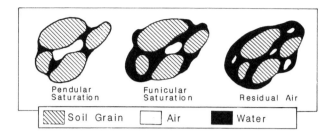

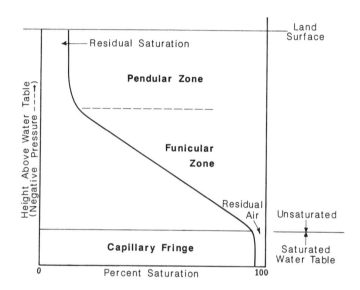

Figure 4.5.   Soil zones defined by water saturation (after Abdul, 1988).

4.4.1.2   *Water Flow Through the Unsaturated Zone*

The primary forces determining the rate of water flow through the unsaturated zone are gravity (acting as a downward force) and a combination of capillary and adhesive forces, called the moisture potential (usually upward). Moisture potential is often also described as the tendency of soil to retain moisture. Depending upon the moisture content, either the gravity force or the retention forces may predominate. When saturation is such that the water phase becomes continuous from pore space to pore space, flow may be possible. At saturations without continuous water connection between pores, the moisture potential is greater and flow does not occur. As the percentage of pore spaces are filled, the volume of flow increases.

Darcy's equation can be used to describe flow in this region, however, the

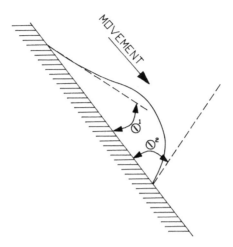

Figure 4.6.   Liquid-solid interface contact showing advancing and re-treating contact angle.

value of permeability varies as a function of saturation. Also, the value of moisture potential is a function of saturation. The total potential for flow (hydraulic gradient in Darcy's equation) can be defined as the difference between the moisture potential (minus) and the elevation potential (plus). When the potential for flow is positive, flow can occur.

Field measurements of moisture potential made by a tensiometer, in conjunction with moisture content, demonstrate that the moisture potential is substantially different if the soil has been in a wetting or drying phase. Soils that have been in a drying phase prior to testing have a greater degree of saturation at the same moisture potential than a soil that has been in a wetting phase. The reason for this hysteresis is threefold: First, as water reenters a dry narrow channel, a local increase in suction is required. If the pore space is too wide, the interface cannot advance until a neighboring pore is filled, allowing the wall-liquid interface angle to be small enough to cause capillary rise. Secondly, the contact angle at an advancing interface differs from that of a receding one, as shown in Figure 4.6. Entrapped air is the third factor causing the hysteresis.

The permeability of wetting or drying unsaturated soils follows a similar relationship with moisture potential as the degree of saturation. A graphic demonstration of the similarity of the two functions is shown in Figure 4.7.

4.4.1.3   Three Phase – Two Immiscible Liquids and Air in the Unsaturated Zone

When a petroleum compound is released into the unsaturated zone, it enters

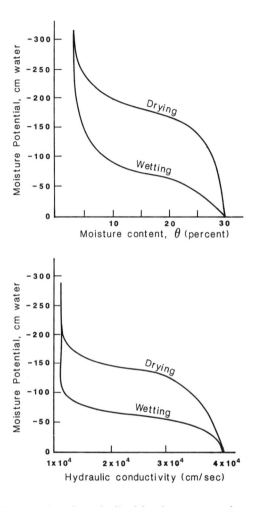

Figure 4.7.   Graphs showing similarities between moisture content vs. moisture potential and hydraulic conductivity vs. moisture potential (after Fetter, 1980).

a very complex environment. The preexisting condition includes partially saturated (with water) pores that are under strong negative pressure, air-filled pores, and a degree of permeability that is variable depending on whether the soil has been recently in a wetting or draining phase. Additionally, this new fluid has its own properties of density, viscosity, and surface tension, all of which influence its ability to flow. In order to move through this environment the LNAPL must satisfy the following:

$$P_{nw} - P_w > \frac{2\sigma\cos\theta}{r}, \tag{4.9}$$

where P = fluid pressure; the subscript nw = nonwetting fluid; the subscript w = wetting fluid; r = interface radius; θ = contact angle; and σ = surface tension when one of the fluids is air or other interfacial tension between two liquid phases.

The above equation states that sufficient pressure must be applied to displace the existing fluid (water or air) before the new fluid can enter the pore spaces. In the pendular zone and the upper part of the funicular zone where residual water does not extend completely from grain to grain, the air offers little resistance. As the LNAPL migrates downward through this region, it tends to flow through interconnected pores, which offer the least resistance, especially the larger pores. LNAPL follows the same rules of physics – wetting, being adsorbed, etc. – as does water. However, most petroleum products are nonpolar and have less strong adhesive attraction to the soil grains, therefore, the "soil suction" is not as significant. In the pendular and upper parts of the funicular zone, gravity is sufficient to overcome the retention potential and the migration is predominantly downward.

The viscosity of separate LNAPL products varies significantly, ranging from far less to many times that of water. Flow of LNAPL in the unsaturated zone is largely dependent upon viscosity and soil grain size. Finer-grained materials have a higher residual saturation of water, which restricts the number of pores available for LNAPL entry in this region.

As more LNAPL enters the soils, it increases the pressure on the moving front, which allows increasing displacement of water. If a sufficient quantity of sufficiently mobile oil is available, it can produce a pressure head large enough to displace water through the funicular zone and to form a distinct interface at the top of the capillary fringe.

A continued thickness increase of LNAPL will depress the water capillary fringe and create a "pancake", with an LNAPL capillary fringe. Laboratory experiments on sandy soils indicate that oil reaches the water capillary fringe region and then moves laterally in the direction of groundwater flow. Higher degrees of oil saturation (>80%) are needed in a sandy medium to displace water from the capillary fringe.

If the quantity of leaked product is not sufficient to overcome the threshold of residual oil saturation, it will be retained and will penetrate only until it is adsorbed by the soil. Estimates of typical oil retention capacities are presented in Table 4.2. This estimate represents porous soils with a "natural" moisture content.

As the petroleum product migrates downward through the unsaturated zone, it will penetrate in a pattern that is characteristic of the soil type and texture. In unstratified homogeneous soil, the form will be pear shaped, with the larger part at the bottom. If the soil is stratified or contains significant fine materials, some horizontal spreading will occur. The general shapes that may be expected are shown in Figure 4.8.

Highly Permeable
Homogeneous Soil

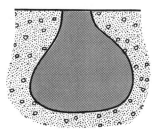

Less Permeable
Homogeneous Soil

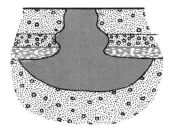

Stratified Soil with
Varying Permeability

Figure 4.8.   **Typical patterns formed by hydrocarbon as it migrates through the unsaturated zone (after CONCAWE, 1979).**

4.4.2   The Saturated Zone (Below the Water Table)

4.4.2.1   Steady State Saturated Flow – Single Fluid

The basic equation for water flow through saturated porous media was developed by Henry Darcy in 1856 to calculate the flow of water through sand filters. This equation has been found to be valid for flow of liquids through porous media when adjusted for the viscosity and density of the liquid. The original form of the equation, designed to calculate discharge, is as follows:

$$Q = \frac{KIA}{\gamma}, \qquad (4.10)$$

where Q = discharge (L^3/T); K = permeability (hydraulic conductivity L/T); I = hydraulic gradient (driving force, or pressure, includes liquid density); γ = viscosity of liquid; and A = cross-sectional area of flow channel L^2.

This form of the equation is valid for liquid flow in the subsurface, when hydraulic conductivity has been determined in relation to water. Common units of hydraulic conductivity are cm/s, ft/d, or m/d.

Although permeability is expressed in units resembling a velocity term, the actual travel rate of the fluid through soil is expressed as:

$$V = \frac{KI}{\phi}, \qquad (4.11)$$

where V = actual travel velocity of fluid through the media; k = hydraulic conductivity (adjusted for viscosity); I = hydraulic gradient (adjusted for liquid density); and ϕ = effective porosity of the media. Addition of the effective porosity (ϕ) to this equation is necessary because the actual cross-sectional area of flow channels through the media is only a fraction of the total area of discharge. The flowing fluid must travel faster through the flow channels to maintain the apparent discharge velocity. Effective porosity describes only those connected pores that are large enough to permit free flow of the fluid. A clay, for example, may have an actual porosity of 50%, however, it may not be capable of transmitting a significant quantity of fluid because the pores are too small or are not connected.

For practical purposes, saturated flow of a single fluid such as gasoline, kerosene, or another particular petroleum product can be predicted by the use of these equations. Standard units of linear measurement (feet, meters, etc.) and discharge are accommodated for by the corrections for viscosity and density. Field testing procedures can be conducted using standard water well testing procedures.

4.4.2.2 Flow of Two Immiscible Fluids

The behavior of mixtures of immiscible fluids flowing through aquifers is often much more important than single-fluid flow during restoration. Recovery

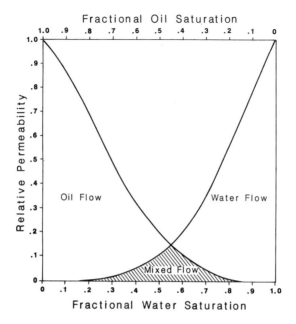

Figure 4.9.   Relative permeability relationship controlling flow of two immiscible fluids (after Leverett, 1939).

of LNAPL usually involves the movement of water through previously oil-saturated soils or oil through previously water-saturated soils. When these situations occur, the migrating fluid is a mixture of the two liquids.

The relative permeability and flow of two immiscible fluids (water and kerosene) in sands was studied by Leverett (1939) and was found to be directly related to fractional water saturation of the aquifer material. The general form of these relationships is shown in Figure 4.9. It should be noted that neither water nor oil is effectively mobile until the relative saturation is in the 20% to 30% range, and even then the relative permeability of the lesser component is approximately 2%.

The mixed flow region of this graph is relatively small in comparison to either the water or oil flow regions. In situations that have nearly equal saturation levels of both water and oil, the relative permeability of each is only approximately 15%. This indicates that greater driving pressures are needed to cause the same total fluid flow quantities through the aquifer under this mixed flow condition.

Each restoration situation would have a specific set of curves to define these relationships. For other materials, the curves may vary in shape, but the concept is similar. Mixed flow substantially reduces the relative permeability. This concept can be very important during the removal of LNAPL by the use of a large drawdown in wells or the raising of water levels for flushing purposes.

4.5   DISPERSION (IN SOLUTION)

Water and hydrocarbons occurring together, in shallow aquifer systems, may be considered "immiscible" for flow calculation purposes, however, each is somewhat soluble in the other. Since groundwater clean-up is the purpose behind restorations, it receives greater attention. Definition of water quality based on samples retrieved from monitoring wells relies heavily upon the concentration of individual chemical components found dissolved in those samples. An understanding of the processes that cause concentration gradients is important for the proper interpretation of analytical results.

The transfer of chemical molecules from oil to water is most often a surface area phenomenon caused by kinetic activity of the molecules. At the interface between the liquids (either static or moving), oil molecules (i.e., benzene, hexane, etc.) have a tendency to disperse from a high concentration (100% oil) to a low concentration (100% water) according to functions of solubility, molecule size, molecule shape, ionic properties, and several other related factors. The rate of the dispersion across this interface boundary is controlled largely by temperature and contact surface area. If the two fluids are static (i.e., no flow), an equilibrium concentration will develop between them and further dispersion across the interface will not occur. This situation is fairly common in the unsaturated zone.

The intensity with which dissolved chemicals are released from liquid hydrocarbon over time is referred to as source strength. Two approaches have been reported that provide an estimate of source strength. Pfannkuch (1984) expressed source strength as mass/time per unit contact area such that:

$$S = K_m A, \tag{4.12}$$

where S = source strength (mass/time/m²); K_m = mass exchange coefficient (mg/m²/s); and A = contact area or interface across which mass exchange occurs (m²). The contact area governs the actual mass that is exchanged within a given period of time. Quantification of the A is very difficult reflecting the complexity of hydrocarbon distribution in the pore space. Thus, efforts to quantify K_m have been attempted. The United States Geological Survey (1984) provided the estimated K_m for certain products as follows:

- Gasoline and tar oil   1.0 mg/m²/s
- Fuel oil, diesel and kerosene   0.01 mg/m²/s
- Lube oils and heavy fuel oil   0.001 mg/m²/s

These mass coefficient values are considered to be maximum values. Field values could be as much as one to two orders of magnitude lower.

The second approach to estimating source strength is also provided by the

United States Geological Survey (1984) and utilizes what is referred to as the Sherwood number and Peclet number:

$$Sh = 0.55 + 0.025\ Pe^{\frac{3}{2}},\qquad\qquad(4.13)$$

where Sh = Sherwood number representing a dimensionless transfer coefficient; and Pe = Peclet number representing a dimensionless flow velocity. When Pe <1, the chemical exchange should be independent of flow velocities and should be essentially diffusion controlled, as anticipated for most soil systems. When Pe >10, the chemical exchange should be dependent upon flow velocities. In addition, if the rate of diffusion for one chemical is known, then the rate for another chemical of similar structure can be estimated as follows:

$$\frac{D_1}{D_2} = \left(\frac{d_2}{d_1}\right)^{0.5} = \left(\frac{M_2}{M_1}\right)^{0.5},\qquad\qquad(4.14)$$

where d_1, d_2 = densities of two chemicals; and M_1, M_2 = molecular weight of two chemicals. The relationship in equation 4.14 is derived from Graham's Law of Diffusion. Generally speaking, the larger the molecule, the slower it diffuses and dissolves. The range of carbon atoms for various petroleum products is shown in Table 3.4.

Below the water table, groundwater is almost always moving at very slow rates. As it travels on its torturous path (Figure 4.10), water molecules mix with other water molecules to dilute the concentration of contaminants within the available pore spaces. This process is referred to as advection. The farther the water travels, the greater the aerial spread of the contaminant.

In conjunction with this mechanical spreading (advection), molecular diffusion continues within the water phase itself. Higher concentrations of dissolved contaminants migrate toward lesser concentrations. The combination of the two diffusing activities typically can be observed by mapping dissolved concentrations of chemical components across a restoration site. An idealized example of dispersion is shown in Figure 4.11. Molecule diffusion can cause an increased concentration in every direction, including the upgradient, especially in slowly moving groundwater. Slower water movement results in more nearly circular diffusion patterns.

The above discussion is most correct when only one contaminating chemical is involved. When contaminants such as gasoline are introduced into the subsurface, the setting becomes more complex. Gasoline is a mixture of mostly small molecules, such as benzene, xylene, toluene, hexane, and other molecules that are mostly less than C_8.

Each individual molecular structure has individual characteristics when dissolving and diffusing through water. Some molecules disperse more easily than others based on the solubility, water-octanol partition coefficient, aquifer

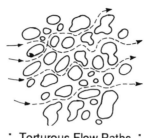

* Torturous Flow Paths *

Figure 4.10.   Torturous flow paths.

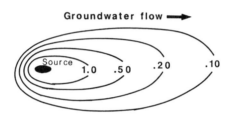

Idealized Dispersion of a Single Component
Indicating Relative Concentrations

Figure 4.11.   Idealized dispersion of a single component indicating relative concentrations.

grain composition, and temperature. The result is a chromatographic distribution of components throughout a dissolved contaminant plume. For gasoline components, benzene disperses more easily than toluene, which spreads more slowly than xylene. In an idealized plume in a uniform aquifer, benzene would be the most widely dispersed of the three chemicals. It would be observed first in downgradient monitoring wells.

Investigators must exercise professional judgment when interpreting analytical results from monitoring wells. In "real world" situations, many other factors are involved that may interfere with the ideal. For example, the soil may contain fine particles (clays or free organic carbon) that have an affinity for a particular component or certain microfluora that biodegrade some molecules more easily than others.

Mechanics of dissolution of LNAPL into water are directly related to surface area contact and the time of contact. Longer contact times raise concentrations (possibly to saturation concentration). Conversely, large droplets of LNAPL dispersed throughout the upper aquifer, with rapidly moving water, may result in much less concentration, even though large quantities of product are present. Based on this discussion, dissolved concentrations in monitoring wells may (or may not) be useful in interpreting the presence of LNAPL.

During the latter phase of remediation of an aquifer, it has often been suggested that the expense of continuing LNAPL recovery is not worth the effort, considering the remaining small quantity of recoverable product. The implication of this argument has been that the product will eventually go into solution and therefore will be recovered as part of the dissolved contaminant recovery. Experience has demonstrated that this is not usually a viable option, because the time required and treatment costs to accomplish the task are much greater with respect to the LNAPL product recovery efforts.

4.6   BIODEGRADATION

Research into the fate of hazardous contaminants has included extensive studies of microbiological activity in the subsurface. Work at contaminated aquifer sites has demonstrated the existence of a wide variety of subsurface microorganisms that exist, are metabolically active, and are often nutritionally diverse. Modern methods of detection have confirmed that it is not uncommon to have uniform population densities of 10^6 to 10^7 cells/g of dry soil in uncontaminated permeable shallow aquifers. The presence of large numbers of microorganisms is dependent upon a number of factors, including surface area available for attachment, oxygen availability, and nutrients to support their growth. Unconsolidated sediments generally have large populations, while subsurface conditions with confining layers or limited permeability will typically have fewer microorganisms.

Many subsurface microorganisms are capable of degrading a variety of contaminants. Typical components of petroleum products (i.e., benzene, toluene, xylene, naphthalene, and simple aliphatic compounds) have been found to be degraded aerobically in unconfined aquifer settings. When petroleum contaminants are introduced into previously uncontaminated soils, a period of acclimation may be required before the indigenous organisms are able to begin the degradation process. This period may extend for a few days to weeks or months, depending upon the variety of organisms present, availability of oxygen, nutrients, concentration of contaminant, and presence of compounds toxic to the microbes.

Both aerobic and anaerobic activity has been reported, however, the majority of case studies are aerobic in nature. Successful continuance of aerobic biodegradation, even in the presence of adequate oxygen, may be limited by insufficient availability of essential nutrients, such as nitrogen and phosphorous, or adverse pH level. Concentrations of the contaminant must be within certain site-specific limits. If the concentration is too low, not enough carbon is available to sustain metabolic activity; conversely, if the concentration is too high, it is often toxic to the microorganisms. A proper balance of organism types, oxygen, nutrient availability, temperature, appropriate soil conditions, and degradable contaminant can result in a rapid and efficient aquifer restoration.

4.6.1   Natural Biodegradation

Most bacteria existing in the subsurface are part of an ecosystem that is low in organic carbon. When these bacteria are cultured in the laboratory with conventional growth media (high concentrations of organic carbon), they do not grow well. These microfluora are adapted to their specific environment. The rate of unenhanced biodegradation of specific contaminants varies significantly. It is not unusual for rate consumption to vary two to three orders of magnitude between aquifers or over a distance of only a few feet (vertically or horizontally). However, even with this variability, aquifer systems are often self decontaminating if enough time is allowed between contamination incidents.

Limitations to the rate of natural biodegradation are related to the total balance of oxygen and the factors discussed previously, but most often the limitation is available oxygen. Biological consumption of benzene requires approximately two parts of oxygen for every part of hydrocarbon. If groundwater has a dissolved oxygen content of 4 ppm, the microorganisms can only degrade 2 ppm benzene before the oxygen is consumed. If the system does not have a ready oxygen recharge capability, the bioclean-up stops. As water solubility of most gasoline components is higher than oxygen solubility, unaided clean-ups cannot proceed until the contaminants are diluted by dispersion and come into contact with oxygen-containing water.

In a few specific instances unplanned bioremediation has been remarkably successful. At these sites, sources of oxygen and nutrients have been available and water table fluctuations have been favorable. The site of a former pumping station was underlain by a shallow alluvial aquifer, with seasonal water table fluctuation of 4–5 ft. At the time of demolition, the aquifer under this site was contaminated with gasoline-type products, both LNAPL and dissolved. After a pumping-recovery system had removed most of the LNAPL phase product, the question of dissolved product removal was debated. During the next 2 years of monitoring, the concentration of dissolved product continually declined more rapidly than the calculated dispersion. An investigation of the processes discovered that seepage from the fertilized wheat field overlying the aquifer provided sufficient excess nitrogen and phosphorous to encourage the bioactivity. Rising and falling of the water table brought the food source (gasoline) to the biomass that was suspended above the water table. Oxygen was readily available due to the high permeability and the shallowness of the water table (undoubtedly, some volatilization of the gasoline occurred).

This ideal, though unplanned, occurrence presents a good example of how some natural environmental systems can remediate themselves, if assisted and not overloaded with new sources of contamination.

4.6.2   Enhanced Biorestoration

The term *enhanced biorestoration* is used to indicate that natural processes are being deliberately manipulated to increase the rate of clean-up. Biological activity in the subsurface is controlled by the availability of one or more of the necessary metabolic requirements discussed previously. When the proper balance of these factors is reached, the rate of chemical consumption is optimized. In most situations the microorganisms are attached to the solid soil particles and await the arrival of water, nutrients, and oxygen. When biomass is above the water table, the dependence is upon the migration of nutrients and diffusion of oxygen downward (or upward through capillary action). Bacterial colonies that develop below the saturated zone are dependent upon liquid phases for the delivery of necessary growth media.

Wilson et al. (1986), summarized the controls of the rate of biological activity as:

1. The stoichiometry of the metabolic process
2. The concentration of the required nutrients in the mobile phases
3. The advective flow of the mobile phases or the steepness of concentration gradients within the phases
4. The opportunity for colonization in the subsurface by metabolically capable organisms
5. The toxicity exhibited by the waste or a cooccurring material

4.6.3   Field Procedures

Procedures have been developed to improve the rate of natural subsurface biological restoration by providing the optimum availability of oxygen and nutrients. This procedure is based on the multiphased approach discussed in the following paragraphs.

A geological and hydrogeological investigation of the site is an important step to evaluate the physical setting for water availability, pore sizes, direction and rate of groundwater flow, and other factors that are needed to assess the "best" enhancement procedure. LNAPL recovery is initiated to remove as much free-phase product as possible. Concurrently, a laboratory analysis of the indigenous microflora is made to determine if it can degrade the contaminant. Further testing of site materials is completed to evaluate the oxygen quantities needed, the proper balance of nutrients, optimum pH, and the presence of toxicants, and to detect other controlling factors. The primary decision is whether to treat the contaminated water above or below ground. If the decision is made to treat the water above ground, it is pumped from wells or sumps treated in above-ground facilities (similar to a wastewater plant) and either returned to flush the aquifer through injection wells (infiltration galleries) or discharged to another disposal outlet. If the water and soil are in the subsurface, plans must be made to provide oxygen and nutrients in sufficient quantities to be effective.

A typical scenario of operation is to strategically place recovery wells at locations that will provide hydraulic containment of the contaminated area while pumping. Injection wells (or infiltration galleries) are located at the edge of the capture zone. Nutrients and oxygen supplied through the injection wells (infiltration galleries) migrate toward the recovery wells (Figure 4.12). Additional wells for oxygen feeding are often installed throughout the site.

Oxygen is usually the most difficult nutrient to supply. Water in the shallow subsurface is normally found to contain between 3 and 4 ppm dissolved oxygen (D.O.). At this concentration, the bioactivity will be limited. Increased oxygen can be provided by air sparging or by the use of hydrogen peroxide.

Air sparging is usually accomplished by forcing air through a porous ceramic block. As the air passes through, it forms very small bubbles which are transmitted into the aquifer and which increase oxygen transfer. When properly installed in wells at the aquifer interface, the air is delivered under pressure, which forces it into the aquifer. Eight to 12 ppm D.O. can be provided by this type of system.

Hydrogen peroxide (H_2O_2) has been found to be a good source of oxygen. It can deliver much higher concentrations than the microflora can tolerate. Often an "inhibitor" is added to control the release monatomic oxygen to a level of <200 ppm near the injection point and near 100 ppm in the zone of bioactivity. At concentrations >100 ppm the H_2O_2 may degas and form bubbles that plug the pore spaces. Iron in the soil can catalyze the decomposition of the H_2O_2. Additional phosphates may be necessary to precipitate the iron to limit its adverse reactions.

Commercial-grade pure oxygen has also been used at remediation sites. Concentrations of 40–50 ppm in the water can be achieved by this method. However, the use of pure oxygen has several disadvantages. Pure oxygen is relatively expensive, may bubble out of solution before the microbes can use it, and may be an explosion hazard if not handled properly. Ozone (O_3) has also been used successfully at remedial projects. Where adequate industrial supplies of this oxidant are commercially available, it provides a good alternative to hydrogen peroxide.

Soil venting is an effective and inexpensive procedure to provide air to the unsaturated zone. When the bulk of the product is held above the water table, supplies of air can be provided by the use of vacuum wells located in this zone. The flow of air is either drawn from the surface to the well (Figure 4.13a) or through vent wells, as shown in Figure 4.13b. The alternative benefit of this approach is that volatile portions of the product are removed by the lower pressure operation. Volatile organic compounds are evaporated from the soil by this method. Soil venting is particularly well suited to less permeable silt and clay soils.

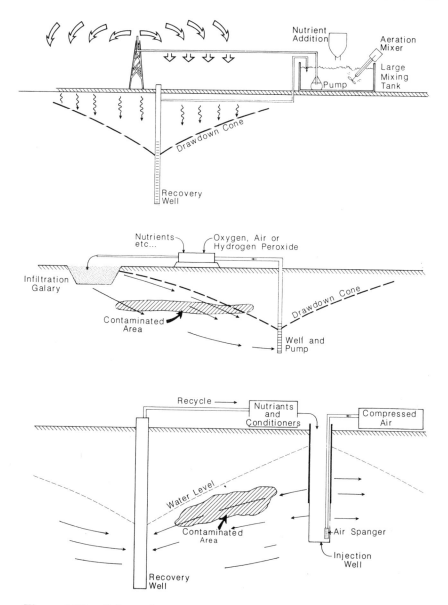

Figure 4.12.   Schematic cross-sections of enhanced biorestoration of aquifers.

Biological treatment of contamination in the subsurface is an effective restoration procedure for many types of organic contamination (especially light petroleum products). Successful use of this technique relies upon optimization of a large number of factors including biological species, soil structure, contaminant concentration, contaminant chemical structure, nutrient balance, subsur-

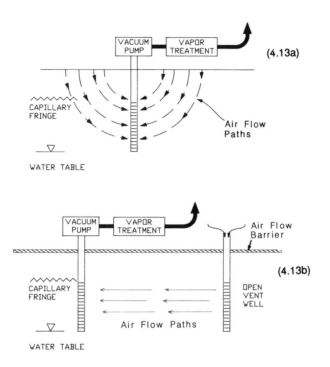

Figure 4.13.   Illustrations showing forced soil venting.

face hydraulic conditions, economic considerations, and regulatory interaction. Because of the complexity of this type of restoration, most effective project teams are composed of a variety of technical specialists, each practicing his or her profession.

REFERENCES

1. Abdul, S. A., 1988, Migration of Petroleum Products Through Sandy Hydrogeologic Systems. *Ground Water Monitoring Review, Fall Issue, 1988,* Vol. VIII, No. 4, p. 73-81.
2. American Petroleum Institute, 1972, The Migration of Petroleum Products in Soil and Groundwater, Principles and Counter-measures: *Am. Petrol. Inst.,* 4149.
3. American Petroleum Institute, Underground Spill Cleanup Manual. *Am. Petrol. Inst.,* 1628, 1st Ed.
4. Bear, J., 1972, *Dynamics of Fluids in Porous Media,* American Elsevier, New York, New York, 764 p.

5. Bear, J., 1979, *Hydraulics of Groundwater:* McGraw-Hill Book Company, New York, New York, 567 p.
6. Bossert, I. and Bartha, R., 1984, The Fate of Petroleum in Soil Ecosystems: *Petroleum Microbiology* (edited by Atlas, R.M.), New York: MacMillan Co., New York, New York.
7. Cherry, J. A., Gillham, R. W., and Barker, J. F., 1983, Contaminants in Groundwater: Chemical Processes: Groundwater Contamination: *Studies in Geophysics,* National Academy Press 1984, p. 46-64.
8. Cherry, J. A., 1984, Groundwater Contamination: in *Proceedings of Mineralogical Association of Canada Short Course in Environmental Geochemistry,* May, 1984, p. 269-306.
9. Chiou, C. T., Peters, L. J ., and Freed, V. H., 1979, A Physical Concept of Soil-Water Equilibria for Nonionic Organic Compounds: *Science,* 206, p. 831-832.
10. CONCAWE Secretariat, 1974, Inland Oil Spill Clean-up Manual, Report No. 4/74. The Hague, Netherlands.
11. CONCAWE, 1979, Protection of Groundwater from Oil Pollution. The Hague, Netherlands, NTIS PB82-174608.
12. CONCAWE, Water Pollution Special Task Force No. 11; Den Haag, 1979. "Protection of Groundwater from Oil Pollution;" Authors: de Pastrovitch, T. L.; Baradat, Y.; Barthel, R.; Chiarelli, A.; Fussell, D. R.
13. Corapcioglu, M. Y. and Baehr, A. L., 1985, Immiscible Contaminant Transport in Soils and Groundwater with an Emphasis on Gasoline Hydrocarbons System of Differential Equations vs. Single Cell Model: *Water Sci. Technol.,* Vol. 17, No. 9, p. 23-37.
14. Corapcioglu, M. Y. and Baehr, A. L., 1987, A Compositional Multiphase Model for Groundwater Contamination by Petroleum Products, 1. Theoretical Considerations: *Water Resour. Res.,* Vol. 23, No. 1, p. 191-200.
15. Dietz, D. N., 1970, Pollution of Permeable Strata by Oil Components: *Water Pollution by Oil* (edited by Hepple, P.), Elsevier Publishing Co. and the Institute of Petroleum, New York.
16. Dragun, J., 1988, *The Soil Chemistry of Hazardous Materials:* Hazardous Materials Control Research Institute, Silver Springs, Maryland, 458 p.
17. Farmer, V. E., Jr., 1983, Behavior of Petroleum Contaminants in an Underground Environment: in *Proceedings of the Petroleum Association for Conservation of the Canadian Environment Seminar on Ground Water and Petroleum Hydrocarbons,* June 26-28, 1983, Toronto, Ontario, Canada.
18. Faust, C. R., Guswa, J. H., and Mercer, J. W., 1989, Simulation of Three-Dimensional Flow of Immiscible Fluids Within and Below the Unsaturated Zone: *Water Resourc. Res.,* Vol. 25, No. 12, p. 2449-2464.

19. Goodwin, M. J., and Gillham, R. W., 1982, Two Devices for In-Situ Measurements of Geochemical Retardation Factors: in *Proceedings of the Second International Hydrogeological Conference* (Ozoray, G., Ed.), International Association of Hydrogeologists, Canadian National Chapter, p. 91-98.

20. Hendricks, D. W., Post, F. J., and Khairnar, D. R., 1979, Adsorption of Bacteria on Soils: Experiments, Thermodynamic Rationale, and Application: *Water, Air, Soil Poll.*, No. 12, p. 219-232.

21. Hoag, G. E., Bruell, C. J., and Marley, M. C., 1984, Study of the Mechanisms Controlling Gasoline Hydrocarbon Partitioning and Transport in Groundwater Systems: Institute of Water Resources, University of Connecticut, United States Department of the Interior Research Project No. G832-06, 51 p.

22. Hunt, J. R., Sitar, N., and Udell, K., 1988, Nonaqueous Phase Liquid Transport and Cleanup, Part I Analysis of Mechanisms. *Water Resourc. Res.*, Vol. 24, No. 8, p. 1247-1258.

23. Karickhoff, S. W., Brown, D. S., and Scott, T. A., 1979, Sorption of Hydrophobic Pollutants on Natural Sediments: *Water Resour. Res.*, Vol. 13, p. 241-248.

24. Kueper, B. H., Abbott, W., and Farquhar, G., 1989, Experimental Observations of Multiphase Flow in Heterogeneous Porous Media: *J. Contam. Hydrol.*, Vol. 5, p. 83-95.

25. Kueper, B. H. and Frind, E. O., 1988, An Overview of Immiscible Fingering in Porous Media: *J. Contam. Hydrol.*, Vol. 2, p. 95-110.

26. Lance, J. C. and Gerba, C. P., 1984, Virus Movement in Soil During Saturated and Unsaturated Flow: *Appl. Environ. Microbiol.*, No. 47, p. 335-337.

27. Lee, M. D., Thomas, J. M., Borden, R. C., Bedient, P. B., Ward, C. H., and Wilson, J. T., 1988, Biorestoration of Aquifers Contaminated with Organic Compounds: *CRC Criti. Rev. Environ. Control*, Vol. 18, Issue 1, p. 29-89.

28. Leo, A. C. Hansch and Elkins, D., 1971, Partition Coefficients and Their Uses: *Chem. Rev.*, No. 71, p. 525-616.

29. Leverett, M. C., 1939, Flow of Oil-Water Mixtures Through Unconsolidated Sands: Transcript of American Institute of Mining and Metallurgical Engineers, Vol. 132, Petroleum Development and Technology, p. 149-171.

30. MacKay, D. M., Cherry, J. A ., Freyerg, D. L., Hopkins, G. D., McCarty, P. L., Reinhard, M., and Roberts, P. V., 1983, Implementation of a Field Experiment on Groundwater Transport of Organic Solutes: in *Proceedings of the National Conference on Environmental Engineering*, ASCE, University of Colorado, Boulder, July, 1983.

31. Maguire, T. F., 1988, Transport of a Non-Aqueous Phase Liquid within a Combined Perched and Water Table Aquifer System: in *Proceedings of the National Water Well Association of Ground Water Scientists and Engineers FOCUS Conference on Eastern Regional Ground Water Issues,* September, 1988.

32. Martin, J. P. and Koerner, R. M., 1984, The Influence of Vadose Zone Conditions on Groundwater Pollution, Part I: Basic Principles and Static Conditions: *J. Hazardous Mater.,* No. 8, p. 349-366.

33. Martin, J. P. and Koerner, R. M., 1984, The Influence of Vadose Zone Conditions in Groundwater Pollution, Part II: Fluid Movement: *J. Hazardous Mater.,* No. 9, p. 181-207.

34. Newsom, J. M., 1985, Transport of Organic Compounds Dissolved in Ground Water: *Ground Water Monitoring Rev.,* Spring Issue, Vol. 5, No. 2, p. 28-36.

35. Pfannkuch, H. O., 1984, Determination of the Contaminant Source Strength from Mass Exchange Processes at the Petroleum-Groundwater Interface in Shallow Aquifer Systems: in *Proceedings of the National Water Well Association and American Petroleum Institute Conference on Petroleum Hydrocarbons and Organic Chemicals in Groundwater—Prevention, Detection, and Restoration,* November, 1984, Houston, Texas.

36. Raisbeck, J. M. and Mohtadi, M. F., 1974, The Environmental Impacts of Oil Spills on Land in the Arctic Regions: *Water, Air, Soil Pollut.,* No. 3, p. 195-208.

37. Roberts, P. V., McCarty, P. L., Reinhard, M., and Schreiner, J., 1980, Organic Contaminant Behavior During Groundwater Recharge: *J. Water Pollut. Control,* p. 161-172.

38. Sawyer, C. N., 1978, *Chemistry for Environmental Engineering Basic Concepts from Physical Chemistry;* McGraw-Hill Book Company, New York, New York.

39. Schwarzenbach, R. P. and Westall, J., 1981, Transport of Nonpolar Organic Compounds from Surface Water to Groundwater, Laboratory Sorption Studies: *Environ. Sci. Technol.,* Vol. 15, p. 1350-1367.

40. Schwille, F., 1984, Migration of Organic Fluids Immiscible with Water in the Unsaturated: Pollutants in Porous Media (edited by Yaron, B., Dagan, G., and Goldshmid, J.), *Ecol. Stud.,* Vol. 47, p. 27-48. Springer-Verlag, New York..

41. Schwille, F., 1985, Migration of Organic Fluids Immiscible with Water in the Unsaturated and Saturated Zones: in *Proceedings of the National Water Well Association Second Canadian/ American Conference on Hydrogeology;* Banff, Alberta, Canada, June, 1985, p. 31-35.

42. Thomas, J. M., Clark, G. L., Tomson, M. B., Bedient, P. B., Rifai, H. S., and Ward, C. H., 1988, Environmental Fate and Attenuation of Gasoline Components in the Subsurface: Rice University, Department of Environmental Science and Engineering, 111 p.

43. United States Geological Survey, 1984, Groundwater Contamination by Crude Oil at the Bemidji, Minnesota, Research Site, U.S.G.S. Toxic Waste-Groundwater Contamination Study: U.S.G.S. *Water Resour. Invest. Rep.,* No. 84-4188.

44. Van Dam, J., 1967, The Migration of Hydrocarbons in a Water Bearing Stratum: *Joint Problems of the Oil and Water Industries* (Hepple, P., Ed.). Institute of Petroleum, London.

45. Van Duijvenbooden, W. and Kooper, W. F., 1981, Effects on Groundwater Flow and Groundwater Quality of a Waste Disposal Site in Noordwijk, The Netherlands: *Sci. Total Environ.,* No. 21, p. 85-92.

46. Vanloocke, R., DeBorger, R., Voets, J. P. and Verstraete, W., 1975, Soil and Groundwater Contamination by Oil Spills; Problems and Remedies: *Int. J. Environ. Stud.,* No. 8, p. 99-111.

47. Villaume, J. F., 1985, Investigations at Sites Contaminated with Dense Non-Aqueous Phase Liquids (NAPL's): *Ground Water Monitoring Rev.,* Spring Issue, Vol. 5, No. 2, p. 60-74.

48. Yang, W. P., 1981, Volatilization, Leaching, and Degradation of Petroleum Oils in Sand and Soil Systems: Ph.D. Thesis, Department of Civil Engineering, North Carolina State University.

5 LNAPL HYDROCARBON DETECTION AND OCCURRENCE

"When in doubt, measure it; when not in doubt, measure it anyway"

5.1 MONITORING WELL INSTALLATION AND DESIGN

Subsurface geologic and hydrogeologic conditions, as well as the subsurface presence of hydrocarbons, can directly be determined by the drilling of borings and the subsequent installation and construction of monitoring wells. Several techniques are available for the drilling and installation of wells, whether their eventual use will be for monitoring, gauging, delineation, injection, or recovery purposes. These techniques are discussed fully in Acker (1974), Campbell and Lehr (1973), and Driscoll (1986). A summary of these techniques is provided in Table 5.1. A typical monitoring well construction detail is shown in Figure 5.1.

Although well construction details for the monitoring and recovery of LNAPL are similar to those of conventional monitoring wells, several factors need to be emphasized. Obviously, the well screen must overlap the mobile hydrocarbon interval and be of sufficient length to account for seasonal fluctuations or changes due to recovery or reinjection influences. Filter pack design can also have a bearing on whether hydrocarbon presence is detected or confirmed. Filter packs must be designed to allow not only mobile hydrocarbon, but also capillary hydrocarbon to migrate into the well. Otherwise, since hydrocarbon in the formation can exist at less than atmospheric pressure, a poorly designed filter pack can result in capillary hydrocarbon being unable to migrate into the well. The end result is a much broader areal extent of subsurface hydrocarbon than is accounted for. Thirdly, well design and construction details among a network

81

Table 5.1   Drilling Techniques for the Construction and Installation of Monitoring, Recovery, and Injection Wells

Drilling Technique	Material	Limitations (in feet)	Depth Well Type (M, R, I)[a]	Remarks
Hand-augered	Unconsolidated	15	M, R	Accurate sampling; difficult in coarse sediments or loose sand; physically demanding; fluid levels easily detected; limitations on diameter of borehole due to friction on auger; depth limitations; inexpensive.
Driven	Unconsolidated	25	M	No sampling capability; quick and easy method to detect and monitor shallow fluid levels.
Hollow-stem auger	Unconsolidated	180	M, R, I	Accurate sampling; diameter limitations; fluid levels easily detected; no drilling fluids necessary; problem in fine sediments with smearing the borehole wall during drilling action that can potentially seal off different zones; continuous sampling available.
Jet	Unconsolidated	200	M	Diameter limitations; difficulty in fluid (LNAPL & H$_2$0) level identification; sampling accuracy limited; problem with produced fluids (hazardous if LNAPL is encountered).
Bucket auger	Unconsolidated	100	R, J	Fast; good samples; can install large diameter well; difficult to control caving.

Method	Formation	Depth	Well type[a]	Comments
Cable tool	Unconsolidated	1000	M, R, I	Accurate sampling; fluid levels easily or Consolidated detected; can be slow.
Hydraulic rotary	Unconsolidated or consolidated	2500+	M, R, I	Fast; collection of accurate samples requires special attention; knowledge of drilling fluids used to minimize plugging of certain formations is critical; good for recovery and injection well construction; continuous coring available; problem with produced fluids if LNAPL is encountered.
Reverse circulation	Unconsolidated or consolidated	2000+	M, R, I	Formation relatively undisturbed compared to other methods; large-diameter holes can be drilled; no drilling mud is usually needed because of the hydraulics associated with reverse method; good for recovery and injection well construction; problem with produced fluids if LNAPL is encountered.
Air rotary	Unconsolidated or consolidated	2000+	M, R, I	Fast; cuttings removal rapid; poor sample quality; depth and diameter limitations; formation not plugged with drilling fluids; dangerous with flammable fluids.
Air-percussion	Unconsolidated or consolidated	2000+	M, R, I	Fast; cuttings removal rapid; good in rotary consolidated formations.

[a] M = Monitoring well.
   R = Recovery well.
   I = Injection well.

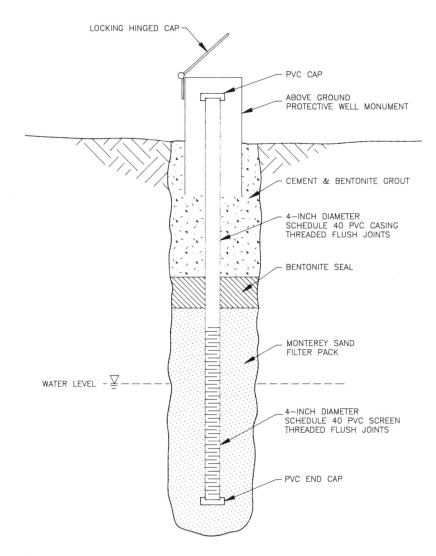

Figure 5.1.   Schematic of typical monitoring well construction detail.

of wells, including the filter pack design and development procedures, should remain consistent. With recovery wells, too coarse a filter pack will minimize the ability of the well to attract capillary hydrocarbon to the well. As Sullivan et al., (1988) clearly state, a typical hydrocarbon product with a density of 0.8 g/cm^3 and an interfacial tension with air of 30 dynes/cm will accumulate to a thickness of approximately 25–33 cm in a fine sand before exceeding atmospheric pressure. A part of this capillary hydrocarbon is recoverable with a properly designed finer grained filter pack.

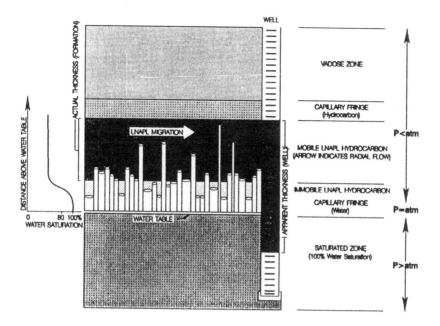

Figure 5.2.   Cause for discrepancies in apparent LNAPL thickness as measured in a monitoring well in comparison with the actual thickness in the formation.

5.2   APPARENT VERSUS ACTUAL THICKNESS

The subsurface presence of LNAPL can occur under both perched and water table conditions (Figure 4.2). In addition, occurrence can exist under unconfined and occasionally confined conditions. LNAPL product in the subsurface is typically delineated and measured by the utilization of groundwater monitoring wells. While monitoring wells have provided some insight as to the extent and general geometry of the plume, as well as the direction of groundwater flow, difficulties persist in determining the actual thickness and, therefore, the volume and ultimately the duration of recovery and remediation. One of the more difficult aspects in dealing with the subsurface presence of hydrocarbons is that accumulations in monitoring wells do not directly correspond to the actual thickness in the formation. The thickness of both LNAPL and DNAPL product as measured in a monitoring well is thus an apparent thickness rather than an actual formation thickness. This relationship is schematically illustrated in Figure 5.2.

Several methods are available to measure the apparent NAPL product thickness in a monitoring well. The apparent thickness of NAPL product in a well is typically determined using either a steel tape with water-and-oil-finding paste or commercially available electronic resistivity probes. Either method can provide data with an accuracy to 0.01 ft. However, if the NAPL product is emulsified or highly viscous, significant error can result. In addition, measurement using electronic resistivity probes can be misleading if the battery source is weak.

The tape-and-paste method involves lowering into a monitoring well a weighted steel measuring tape on which hydrocarbon- and water-sensitive pastes have been applied. As the water and liquid hydrocarbons contact the pastes, color changes occur. The area of sharply contrasting colors made on the paste is called the water cut. The water cut delineates the interface between the floating hydrocarbons and the uppermost surface of groundwater in the well. The top surface of the liquid hydrocarbon is determined by locating the top of the characteristic oily film on the steel tape. The distance between these two markers is recorded from the measuring tape, and the resultant length is equivalent to the apparent thickness of liquid hydrocarbon present in the well.

The interface gauging probe incorporates a measuring tape on a reel, connected to an electronic sensor head. The sensor head contains a floatball and magnetic relay switch assembly that distinguishes between air and fluids; also contained in the sensor head is an electrical conductivity sensor. This sensor consists of two electrodes, between which a small electrical current is passed. This electrical conductivity sensor distinguishes between nonconductive fluids (hydrocarbons) and conductive fluids (groundwater).

The sensor head is lowered into a monitoring well. Upon contact with any fluid the float ball is raised and a continuous tone is emitted from an audible alarm. When the sensor head contacts the interface between LNAPL and groundwater, the change in conductive properties is detected by the electrical conductivity sensor and a beeping tone is emitted. The distances along the tape at which the two changes in the audible alarm occur are recorded as referenced from a presurveyed point on the lip of the monitoring well. The resultant distance is equivalent to the apparent thickness of the LNAPL in the well.

The third method, although not typically used, is a transparent bailer that is lowered into the well. A bailer is a cylindrical device with a check valve on the bottom and a hook for a cord on the top. The bailer must be long enough to ensure that its top will be above the air/LNAPL interface when the check valve is below the LNAPL/water interface. The hydrocarbon thickness measured in a bailer can be slightly greater than that actually present in the well. This is because a volume equivalent to the bailer wall thickness will displace the hydrocarbons. Some of these displaced mobile hydrocarbons will enter the bailer, thus exaggerating the apparent thickness of the LNAPL actually present. The bailer should thus be lowered slowly through the hydrocarbon layer to minimize this discrepancy.

Since hydrocarbon and water are immiscible fluids, the hydrocarbons are perched on the capillary fringe above the actual water table. The physical relationships that exist are illustrated in Figure 5.2. This discrepancy can be a

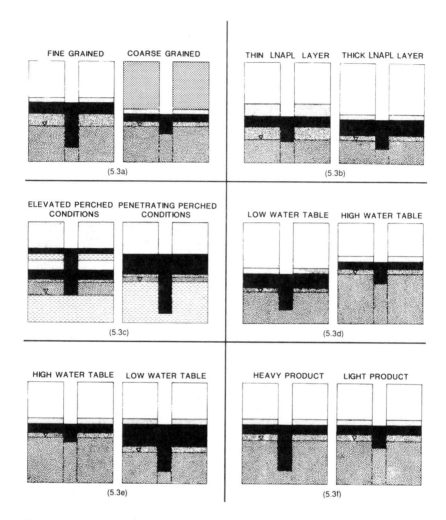

Figure 5.3. Schematic showing generalized relationship of actual vs. apparent LNAPL thickness in the well and adjacent formation.

result of one or combination of factors or phenomenon. Some of the more common factors or phenomenon are schematically shown in Figure 5.3 and include

- Grain size differences reflected in varying heights of the capillary fringe
- Actual formation thickness of the mobile LNAPL hydrocarbon

Table 5.2   General Capillary Rise for Certain Soil Types

Soil Type	Capillary Rise (in.)
Coarse sand	3/4–2
Sand	4–14
Fine sand	14–27
Silt	27–59
Clay	7–160+

Source: From Bear (1979).

• Height of perching layers, if present
• Seasonal or induced fluctuations in the level of the water table
• Product types and respective specific gravities
• Confining conditions

The capillary fringe height is grain size dependent, as summarized in Table 5.2 and shown in Figure 5.3a. As grain size decreases, the capillary height increases. Coarse-grained formations contain large pore spaces that greatly reduce the height of the capillary rise. Fine-grained formations have much smaller pore spaces which allow a greater capillary height.

Since the LNAPL product occurs within and above the capillary fringe, once the bore hole or monitoring well penetrates and destroys this capillary fringe, LNAPL product migrates into the well bore. The free water surface that stabilizes in the well will be lower than the top of the surrounding capillary fringe in the formation, thus, mobile hydrocarbons will flow into the well from this elevated position. Product will continue to flow into the well and depress the water surface until a density equilibrium is established. To maintain equilibrium, the weight of the column of hydrocarbon will depress the water level in the well bore. Therefore, a greater apparent thickness is measured than actually exists in the formation, thus the measured or "apparent" LNAPL thickness is greater for fine-grained formations and less for coarser-grained formations, which may be more representative of the true thickness.

The measured or "apparent" hydrocarbon thickness is not only dependent upon the capillary fringe, but also on the actual hydrocarbon thickness in the formation (Figure 5.3b). In areas of relatively thin LNAPL accumulations, the error between the apparent well thickness and the actual formation thickness can be more pronounced than in areas of thicker accumulations. The larger error reflects the relative difference between the thin layer of LNAPL in the formation and the height it is perched above the water table. The perched height is constant

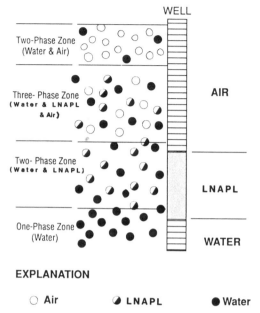

Figure 5.4.   Conceptual representation of distribution of air, LNAPL, and water in a porous medium and monitoring well.

for thick and thin accumulations; however, a thick accumulation can depress and even destroy the capillary fringe, as illustrated in Figure 4.2.

A conceptual illustration of the distribution of water, LNAPL, and air in a porous media is discussed by Farr et al., (1990) and is presented in Figure 5.4. In Figure 5.4, the continuous pore volume is occupied by one of three fluids:

• A two-phase zone containing water and air
• A three-phase zone containing water, LNAPL and air
• A two-phase zone containing water and LNAPL
• A one-phase zone containing solely water

The formation of a distinct LNAPL layer floating on top of the capillary fringe, as illustrated in Figure 5.3, would violate the fundamental equations that describe the fluid pressure distributions in a porous media and also in the monitoring well under conditions of mechanical equilibrium (Farr et al., 1990). This accounts in part for the poor LNAPL yields from spill sites, thus, the presence of LNAPL overlying the capillary fringe as shown in Figure 5.2 and observed in the laboratory, may reflect entrapped air plus water, which over time would be released from this zone providing continuous pore space.

The thickness measured in a monitoring well with LNAPL product situated on a perched layer at some elevation above the water table, can produce an even larger associated thickness error (Figure 5.3c). This commonly occurs when the well penetrates the perched layer and is screened from the perching layer to the water table. LNAPL then flows into the well from the higher or perched elevation. The accumulated apparent thickness is a direct result of the difference in their respective heights. If a situation such as this exists, the difference in the respective heights and weights of the column of hydrocarbon should be accounted for in determining the actual thickness.

Additionally, vertical fluctuations in the water table due to recovery operations or seasonal variations in precipitation have a direct effect upon the apparent or measured LNAPL thickness (Figure 5.3d). As the water table elevation declines gradually due to seasonal variations, for instance, an exaggerated apparent thickness occurs, reflecting the additional hydrocarbon that accumulated in the monitoring well. The same is true for an area undergoing recovery operations where the groundwater elevation is lowered through pumping, and thicker apparent thicknesses may be observed.

The reverse of this effect has also been documented at recovery sites (Figure 5.3e). When sufficient recharge to the groundwater system through seasonal precipitation events or cessation of recovery well pumping occurs with the water table at a slightly higher elevation, thinner LNAPL thicknesses may be observed. During this situation a compression of the capillary zone occurs, lessening the elevation difference between the water table and the free hydrocarbon, which reduces the apparent thickness.

Differences in product types, and thus in API gravities, can account for variations in the apparent thickness as measured in a monitoring well (Figure 5.3f). Heavier product types of relatively low API gravities will tend to depress the free hydrocarbon/water level interface more than would be anticipated from a product of higher API gravity.

In addition, monitoring wells screened across LNAPL within confined aquifers will exhibit an exaggerated thickness. This exaggerated thickness reflects relatively high confining pressures, which force the relatively lower density fluid upward within the borehole. Thus, the measured thickness is a function of the hydrostatic head and not the capillary fringe, which has been destroyed by the confining pressures.

It is often assumed that apparent thicknesses are greater near the edge of a LNAPL pool and smaller toward the center. The relationship between apparent thicknesses as measured in wells and formation thickness across a LNAPL pool is shown in Figure 5.5. Ratios are essentially smaller near the center of the pool, where the thickness of the LNAPL is sufficient to displace water from the capillary fringe. Conversely, where the LNAPL thickness is less, toward the edge of the pool, ratios increase.

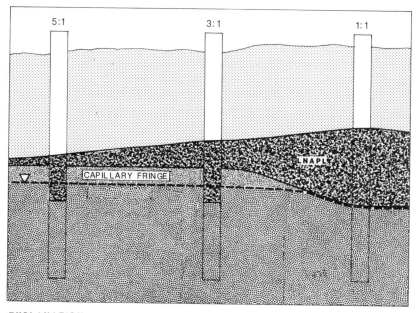

EXPLANATION

Mobile LNAPL

Water Table

Figure 5.5.   Cross-section of a LNAPL pool showing well to formation LNAPL thickness ratios.

5.3   DETERMINATION OF APPARENT VERSUS ACTUAL LNAPL THICKNESS

Various approaches and techniques have been used for the determination of the actual thickness of LNAPL in the subsurface. These approaches essentially fall into two groups: indirect empirical and direct field approaches. Both of these are further discussed in the following subsections.

5.3.1   Indirect Empirical Approach

Several equations and empirical relationships have been formulated in an attempt to relate the apparent product thickness as measured in a monitoring well to that which actually exists in the adjacent formation. These indirect empirical approaches are summarized in Table 5.3 and are discussed below.

Van Dam (1967) initially provided the theory and understanding required to quantitatively deal with spills and illustrated the physical processes responsible for product accumulation in wells and adjacent porous media. Zilliox and Muntzer (1975) evaluated the relationship between actual and apparent thicknesses using a physical laboratory model and proposed the following equation:

$$\Delta h = H - h = \frac{P_c^{wo}}{(\rho_w - \rho_o)g} - \frac{P_c^{oA}}{(\rho_o - \rho_A)g}, \qquad (5.1)$$

where H = apparent product thickness, measured in well; h = average product thickness in soil near well; Δh = difference in thicknesses; P_c^{wo} = pressure difference (capillary pressure) between water and oil at their interface; P_c^{oA} = capillary pressure between oil and air; ρ_w, ρ_o, ρ_A = density of water, oil, and air; and g = gravitational acceleration.

These variables are illustrated in Figure 5.6. Zilliox and Muntzer evaluated the effects of falling and rising water tables on Δh in their plexiglass laboratory model. When the water table fell, the capillary pressures P_c^{wo} and P_c^{oA} were nearly equal and Δh was positive at equilibrium. When the water table rose, they observed that P_c^{wo} varied little and P_c^{oA} decreased considerably. As a result, Δh decreased and become negative. Hence, the actual thickness could be greater than the apparent thickness.

De Pastrovich et al., hereafter referred to as CONCAWE (1979), presented a modified version of Zilliox and Muntzer's equation, as illustrated in Figure 5.6 and presented below:

$$\frac{H}{h} \approx \frac{H-a}{h-a} = \frac{P_c^{wo}}{P_c^{oA}} \frac{(\rho_o - \rho_a)g}{(\rho_w - \rho_o)g} = \frac{P_c^{wo}}{P_c^{oA}} \frac{\rho_o}{(\rho_w - \rho_o)} \approx 4 \frac{P_c^{wo}}{P_c^{oA}} \approx 4, \qquad (5.2)$$

where h = actual product thickness including product fringe; and a = product-saturated portion of h.

In equation 5.2, the factor 4 results from assuming that $\rho_o = 0.8\ \rho_w$. This ratio $\rho_o/(\rho_w - \rho_o)$, hereafter called the "CONCAWE factor," actually varies from 2 to 24 as ρ_o varies from 0.67 ρ_w to 0.96 ρ_w (Table 5.4). CONCAWE (1979) suggested that "more often than not $P_c^{wo} \approx P_c^{oA}$ from which follows that H may be roughly four times h." According to Zilliox and Muntzer, this assumption of equal capillary pressures applies when the water table falls. If all of the CONCAWE assumptions are correct except $\rho_o = 0.8\ \rho_w$, then the CONCAWE factor should equal the ratio H/h.

Revisions to this well-known but poorly understood rule of thumb have been suggested. Kramer (1982), after citing CONCAWE (1979) in discussing a gasoline spill, pointed out that "field experiences have shown that the product thickness in monitoring wells represents two to three times the thickness in the formation." At another gasoline spill into fine glacial sands, Yaniga (1982) observed that the apparent thickness was $2^1/_2$ to 3 times greater than the real thickness. For a #2 fuel oil spill at one particular site, Yaniga and Demko (1983) claimed that "apparent product thicknesses proved to be five to ten times greater than representative 'actual' values." These figures are actually consistent with equation 5.2 if the CONCAWE factor of 4 is corrected for product density, as in Table 5.4. The comparison is shown in Table 5.5.

Equation 5.2 has been disputed by several investigators. Shepherd (1983) claimed that apparent product thicknesses in wells may become more exaggerated than accounted for in equation 5.2 in certain cases. Hall et al., (1984) described convincing laboratory experiments that contradicted equation 5.2. Their results can be formalized as follows:

$$h = H - F, \tag{5.3}$$

where h = actual product thickness; H = apparent product thickness, greater than minimum value (Table 5.6); and F = formation factor (Table 5.6).

Note that F is a function of grain size and that it follows the trend of capillary fringe height for different sand sizes. Equation 5.3 applies only to sands in which H exceeds a minimum value which is also related to capillary fringe height.

For cases where an aquifer is believed to be contaminated by a product spill but no apparent product thickness measurements are available, Dietz (1971) presented Table 5.7 relating the maximum expected product thickness to the average aquifer grain diameter for various sands. He assumed that the maximum oil pancake thickness equals the capillary zone thickness. For each sand size range, he gave a corresponding range of maximum h values.

Schiegg (1985) conducted exquisite laboratory experiments in this area and developed another equation for the actual thickness:

$$h = H - 2 \, \overline{h}_{c,dr}, \tag{5.4}$$

where $\overline{h}_{c,dr}$ = mean capillary height of saturation curve for water and air during drainage; and $\overline{h}_{c,dr} \approx$ height of the capillary fringe during drainage.

In order to use equation 5.4, one must measure or calculate $h_{c,dr}$ for a given aquifer material. A mathematical formula for the water pressure-saturation curve during drainage is needed to calculate the average water pressure head. The Brooks-Corey model of the drainage curve can be used as follows (Corey, 1986):

$$S_e = \frac{S - S_r}{S_m - S_r} = \left(\frac{h_d^\lambda}{h_c}\right) \text{ or } h_c = h_d \, S_e^{-\frac{1}{\lambda}} \tag{5.5}$$

Table 5.3 Summary of Equations Relating Apparent LNAPL Thickness (H) to Actual LNAPL Thickness (h)

Equation	Actual LNAPL Thickness as a Function of			Reference
	Aquifer Properties?	Apparent LNAPL Thickness (H)	Oil Density?	
$h = Z_L - Z\mu$	Yes?	Yes?	Yes?	van Dam (1967)
if $S_o > 0.2$ for $Z_L > Z > Z\mu$ $h_{max} = f$ (aquifer grain size)	Yes	No	No	OILEQUIL Dietz (1971)
$h = H + \dfrac{P_c^{OA}}{(\rho_o - \rho_A)g} - \dfrac{P_c^{OW}}{(\rho_w - \rho_o)g}$	Yes	Yes	Yes	Zillox & Hunter (1975)
$h = H\left(\dfrac{\rho_w - \rho_o}{\rho_o}\right)\dfrac{P_c^{OA}}{P_c^{OW}} \approx \dfrac{H}{4}$	No	Yes	If adjusted	CONCAWE (1979)
$h = H - F$	Yes	Yes	No	Hall et al. (1984)
$h = H - \overline{2}h_{c,dr} = H - \dfrac{2\lambda h_d}{\lambda - 1}$	Yes	Yes	No	Schlegg (1965)

$h = Z_L (S_o = 0.29) - Z_u (S_o = 0.09)$	Yes	Yes	No	Wilson et al. (1988) OILEQUIL
$h = \dfrac{1}{S_o} \displaystyle\int_0^{Z_{ow}} S_o \, dz$	Yes	Yes	Yes	Parker and Lenhard (1988) QUILEQUIL

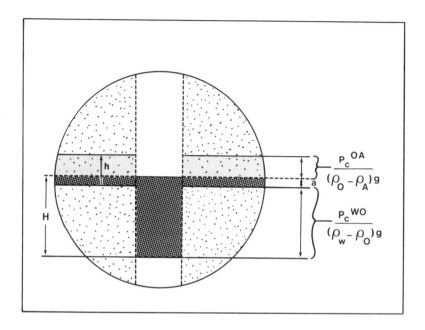

Figure 5.6.   Oil thickness in a well and adjacent formation (from CON-CAWE, 1979).

where S_e = effective saturation; S = saturation = phase volume/pore volume; S_r = residual saturation; S_m = maximum field saturation; h_c = capillary pressure head; h_d = air-entry displacement head; and λ = pore-size distribution index.

To find the average capillary head, equation 5.5 is inserted into the definition of the average head:

$$\bar{h}_{c,dr} = \frac{\int_0^1 h_c dS_e}{\int_0^1 dS_e} = \frac{h_d \int_0^1 S_e^{\frac{-1}{\lambda}} dS_e}{1}$$

$$= h_d \frac{\left[S_e^{-\frac{1}{\lambda}+1} \right]_0^1}{-\frac{1}{\lambda}+1} = \frac{h_d}{\frac{\lambda-1}{\lambda}}(1-0) = \frac{\lambda h_d}{\lambda-1}. \tag{5.6}$$

Now that $\bar{h}_{c,dr}$ is expressed solely in terms of the Brooks-Corey parameters for a given soil, equation 5.4 becomes

$$h = H - 2\,\bar{h}_{c,dr} = H - \frac{2\lambda h_d}{\lambda-1}. \tag{5.7}$$

Table 5.4   CONCAWE Factor Ranges for Various Petroleum Products[a]

Similar Product Types	Product Density, ρ_o	(°API)	(g/cm)[c]	CONCAWE Factor $\dfrac{\rho_o}{\rho_w - \rho_o}$
Gasoline[b]	High	49.6	.781	3.6
	Low	72.1	.695	2.3
	Average	61.1	.735	2.8
JP-4[c]	High	45.7	.800	4.0
	Low	56.7	.752	3.0
	Average	53.9	.763	3.2
#1 D[d]	High	37.8	.836	5.1
#1 fuel oil[e]	Low	47.9	.790	3.8
or kerosene	Average	42.4	.814	4.4
Jet A[c]	High	37.4	.838	5.2
(JP-5)	Low	47.4	.791	3.8
(kerosene)	Average	42.5	.813	4.4
#2 D[d]	High	22.1	.920	11.5
#2 fuel oil[e]	Low	44.5	.804	4.1
or diesel	Average	34.6	.852	5.8

[a]   All data come from various topical reports by the National Institute for Petroleum and Energy Research (NIPER), P. O. Box 2128, Bartlesville, OK 74005. See also Gruse (1967).

[b]   From NIPER motor gasoline surveys for the summers of 1980 and 1981 and for the winters of 1980-1981, 1981-1982 and 1985-1986.

[c]   From NIPER aviation turbine fuels survey in 1981.

[d]   From NIPER diesel fuel survey in 1981.

[e]   From NIPER heating oil surveys in 1981, 1982 and 1986.

Table 5.5   Published Values for H/h

Spilled Product Type	Reported H /h	CONCAWE Factor Range	Reference
Gasoline	2–3	2.3–3.5	Kramer (1982)
Gasoline	2.5–3	2.3–3.5	Yaniga (1982)
#2 Fuel oil	5–10	4.0–11.5	Yaniga & Demko (1983)

Table 5.6   Information Required to Apply Equation 5.3

Classification[a]	Minimum Apparent Thickness	F, Formation Factor	Capillary Fringe Height[b]	U.S. Standard Sieve Size Range
Coarse sand	8 cm	5 cm	2–5 cm	#4–#10
Medium sand	15 cm	7.5 cm	12–35 cm	#10–#40
Fine sand	23 cm	12.5 cm	35–70 cm	#40–#200

[a] Based upon U. S. Bureau of Reclamation scale using median grain size.
[b] From Bear (1979, p. 75).

Table 5.7   Estimate of LNAPL from Aquifer Grain Size[a]

Sand Size	Average Grain Diameter, mm[b]	Corresponding Capillary Zone Thickness, cm[c]
Extremely coarse to Very coarse	2.0–0.5	1.8–9.0
Very coarse to Moderately coarse	0.5–0.2	9.0–22.4
Moderately coarse to Moderately fine	0.2–0.05	22.4–28.1
Moderately fine to Very fine	0.05–0.015	28.1–45.0

[a] From Dietz (1981).
[b] Maximum actual oil pancake thickness, h.
[c] Function of aquifer average grain size.

Parker et al. (1987) have developed constitutive relations describing the various multiphase saturation curves. With these functions, one can then apply the quantitative analyses of van Dam (1967) to a given soil.

Parker and Lenhard (1989) and Lenhard and Parker (1989) have developed equations that relate the apparent product thickness measured at a well under equilibrium conditions with the product and water saturations in a vertical column of soils adjacent to the well. By integrating the product saturation curve with respect to elevation, an equivalent depth of product-saturated pores is obtained. This process has been implemented in a computer program called OILEQUIL. The result is reported as a total oil depth in a vertical profile. The water and oil saturation curves with elevation can also be produced and printed in graphical or tabular form.

The results from OILEQUIL can be converted to an active layer thickness in several ways. First, the total thickness of the zone shown to contain product could be used. This would overestimate the thickness of the recoverable product

layer, because oil at low saturations is immobile. Second, one could follow van
Dam (1967) in assuming a residual oil saturation of 20% (or another value
appropriate to the soil and product involved) and pick out the thickness of the
zone with greater than 20% product saturation from tabular output of OILE-
QUIL. In equation form,

$$h = Z_L - Z_u \text{ and } S_o > 0.2 \text{ for } Z_L > Z > Z_u, \tag{5.8}$$

where h = actual product thickness; Z = vertical coordinate, positive downward; S_o = oil saturation; Z_L = depth to lower boundary at which S_o = 0.2; and Z_u
= depth to upper boundary at which S_o = 0.2.
    Another similar approach would be to follow Wilson et al. (1988) in using
different residual saturations for the vadose and water-saturated zones. Wilson
et al. (1988) found s_r values of 9% and 29% in the vadose and saturated zones,
respectively. Hence, one could revise equation 5.8 as follows:

$$h = Z_L(S_o = 0.29) - Z_u(S_o = 0.09). \tag{5.9}$$

This approach has promise, particularly if one can determine the residual
product saturation in an independent laboratory experiment. This experiment
would consist of acquiring an "undisturbed" sample of the aquifer just above the
water table, wetting it with water and allowing it to drain for a few days,
saturating it with water and allowing it to drain for a few days, saturating it with
product and allowing it to drain for a week or two. Then the product mass and
dry soil bulk density would be determined via freon extraction and soil drying,
respectively. Using these quantities, one could calculate the residual product
saturation.
    Third, the total oil thickness (or thickness of saturated pores) determined by
OILEQUIL could be converted to an active layer thickness using an assumed
average oil saturation for the active product layer. One would divide the total oil
thickness by the expected average saturation to get the actual thickness. Hence,
a 2 cm total oil thickness would be equivalent to a 3 cm thick active product layer
with an average saturation of 66.7%. In equation form,

$$h = \frac{1}{\overline{S_o}} \int_0^{Z_{ow}} S_o dz, \tag{5.10}$$

where $\overline{S_o}$ = average oil saturation of active layer; z = depth below land surface;
Z_{ow} = oil/water interface depth in monitoring well; and o = land surface datum.
    Table 5.3 summarizes the equations presented above. Each incorporates
various assumptions. These assumptions limit the ability of the equations to
account for all of the necessary physical processes. Any equation that does not
incorporate all of the important physical processes cannot work, in general. The

equations can further be evaluated for the three major factors: measured aquifer properties, apparent thickness, and product density (Table 5.3). The CON-CAWE equation is the only one to omit the use of aquifer material properties. Since no soil or rock characterization is required, this method is often used. However, one would not anticipate the same product thicknesses to be found in sand, silt, and clay aquifers for a given apparent thickness. The method of Parker and Lenhard (1989) appears to be the only one amenable to analysis that includes all three major components. Other physical components not shown in Table 5.3 include saturation history, hysteresis, and interfacial tensions.

Superimposed on the difficulty in estimating the actual thickness using product thickness measurements in monitoring wells is the influence of well diameter on such measurements. Chaffee and Weimar (1983) reported that the gasoline thicknesses ranged from 6 to 30 in. in a small-diameter monitoring well and from 1 to 2 in. in a large-diameter well nearby. Similarly, Cummings and Twenter (1986) found 1.27 ft. of JP-4 in a 3-in. diameter well and 0.01 ft. of fuel in a 6-in. well about 4 ft. away. They deepened and pumped the 6-in. well, but did not significantly reduce fuel thickness in the 2-in. well. Mansure and Fouse (1984) explain these observations using thickness data from a monitoring-well network that included wells at the edge of the gravel pack around large-diameter recovery wells. They convincingly demonstrated the difficulty with which fuel migrates from an aquifer into a large-diameter, gravel-packed well. During pumping, product accumulates outside of the gravel pack. Review of laboratory results have shown that none of these methods can be trusted without further validation. Although some have provided reasonable estimates, cross-validation and field validation is greatly insufficient.

5.3.2   Direct Field Approach

Several direct field approaches have been reported for the measurement of actual LNAPL product thickness. These approaches are summarized in Table 5.8 and discussed further below.

Continuous dry coring (without water or mud) of the interval immediately above and below the mobile LNAPL layer seems to be an obvious way of determining the mobile LNAPL thickness. Standard split-spoon samples are commonly used for this purpose. The use of clear acrylic Shelby tubes have also been used with favorable results. However, several problems are associated with coring which results in overestimating the actual thickness of the product. These factors include:

- Loss of material each time the sample is driven then extracted
- Loss of sample in saturated or partially saturated unconsolidated sediments during extraction

Table 5.8   Direct Field Methods for Measuring Actual LNAPL
Hydrocarbon Thickness

Method	Reference
Bailer test	Yaniga (1982)
	Yaniga and Demko (1983)
Continuous core analysis	Yaniga (1984)
Continuous core analysis with	Hughes et al. (1988)
ultraviolet light (natural fluorescence)	
Test pit	Hughes et al. (1988)
Bail-down test	Gruszczenski (1987)
	Hughes et al. (1988)
Recovery well recharge test	Hughes et al. (1988)
Dielectric well logging tool	Keech (1988)
Optoelectronic sensor	Kimberlin and Trimmell (1988)

- Depth control
- Nonrepeatable
- Increased drilling costs

Fluorescence can be used in conjunction with coring and provide a qualitative tool in evaluating the presence of hydrocarbons and the relative abundance of fluorescent hydrocarbons present in the affected zone. Most hydrocarbons fluoresce including light-colored gasolines and jet fuels. Even though these types of hydrocarbons may not be detectable under white light, ultraviolet (UV) light can be used to detect their presence. Under UV light, samples whose pore space is filled with air or water will remain dark, while samples whose pore space is filled with hydrocarbon will glow. All other things being equal, the higher percentage of hydrocarbon in total volume will glow the brightest.

The test pit method consists of excavating a pit using a backhoe to a depth where hydrocarbon pools in the bottom of the pit. Soil samples are subsequently obtained above and below the hydrocarbon-affected zone using hand-driven tubes (acrylic small diameter Shelby tubes). Problems associated with this method include:

- Impracticality of excavating test pits due to site-specific factors (i.e., piping, utilities, etc.)
- Coring in unconsolidated sediments
- Generation of soils that may be considered hazardous and require special handling

Bail-down testing is a widely used field method to evaluate the "actual" thickness of LNAPL product in a monitoring well. Bail-down testing was originally used as a field check method to determine potential locations for LNAPL recovery wells. All monitoring wells at a site that exhibited a measurable thickness of LNAPL product were typically tested. Whether or not all the LNAPL product could be removed from the well and the volume of liquid hydrocarbon bailed were general indicators of areas for "potentially good" recovery.

Under certain circumstances, the actual thickness of the mobile LNAPL layer in the formation can be determined by conducting a bail-down test. Bail-down tests involve the estimation of actual thickness via the graphical presentation of depth-to-product, depth-to-water, and apparent product thickness vs. time as measured during the fluid recovery period in each well as shown in Figure 5.7. This procedure is useful when the mobile LNAPL/water interface is above the potentiometric surface, the rate of mobile LNAPL into the well is slow, and sufficient LNAPL thickness is present (greater than 0.1 in.). Bail-down testing field procedures are similar to those performed for *in situ* permeability tests and involves the measurement of the initial "apparent" thickness in the monitoring well by an oil/water interface gauging probe or pressure transducer. Two different procedures have been proposed (Gruszczenski, 1987 and Hughes et al., 1988). With Gruszczenski's (1987) method, both water and product is bailed from the well until all of the hydrocarbon is removed or no further reduction in thickness can be achieved. Measured over time are levels of both depth to product (DTP) and depth to water (DTW). Typically, the time increments for measurement follow the same sequence as monitored during an *in situ* rising or falling head permeability test. The test is considered complete when the well levels have stabilized for three consecutive readings or if a significant amount of time has elapsed and the levels have reached 90% of the original measurements.

According to Gruszczenski (1987), if the apparent thickness is greater than the actual thickness, and the thickness in the well has been reduced to less than actual during bailing, then at some point during fluid recovery the apparent thickness equals the actual thickness. During recovery of the fluid levels in the well, the top of the product in the well rises to its original level. However, the top of water (product/water interface) initially rises and then falls. The fall is due to displacement of the water in the well, reflecting an overaccumulation of product on the water surface. The point at which the depth-to-water graph changes from a positive to negative slope is referred to as the *inflection point*. The distance between the inflection point and the measured stabilized top of product is considered by Gruszczenski (1987) as the actual mobile LNAPL thickness in the formation, as shown in Figure 5.7. The difference between the actual product thickness and the static depth to product prior to testing is the height of the capillary fringe. Where no capillary fringe exists, the distance reflects the actual thickness.

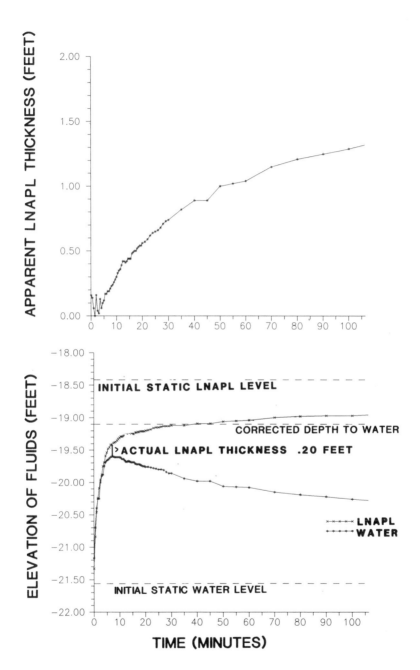

Figure 5.7.   Representative bail-down testing curve results using Gruszczenski's (1987) method. Graphs of depth to product and depth to the product/water interface vs. time (a), and product thickness vs. time (b) are produced.

Another approach to bail-down testing is that proposed by Hughes et al., (1988). According to this procedure, only hydro-carbon (no water) is removed from the well. Hydrocarbon, once removed, will recharge at a constant rate and then will begin to steadily decrease as the well continues to fill. Hydrocarbon removal is accomplished by the use of a bailer or skimmer pump. As the rate of recharge changes, this change reflects not the elevation of the mobile LNAPL/water interface, but the base of the mobile LNAPL in the formation. The actual thickness of the mobile LNAPL in the formation is thus the distance between the point of initial rate change and the static depth to the top of the hydrocarbon layer prior to testing. This relationship can be shown on a graph of the top of a mobile hydrocarbon layer as measured in the well vs. time, as presented in Figure 5.8.

Bail-down test results do not always conform to the theoretical response anticipated. This is evident from Figure 5.7, which shows much fluctuation as a result of bore-hole effects. In these instances, maximum theoretical values can be determined by subtracting the static depth to product from the corrected depth to water. Thicknesses provided in this manner are conservative in that true thicknesses must be less than or equal to these values, and thus overestimates the actual thickness by an amount equal to the thickness of the capillary zone.

Although bail-down testing is a relatively simple field procedure, the analysis and evaluation of the data remains speculative. The method contains a number of steps where errors can easily be introduced. Bail-down testing results are, however, relied upon to determine "actual" thickness in a monitoring well and is an initial step and basis for calculating a volume and ultimately a recoverable volume. Bail-down tests do not provide usable curves for determining actual thickness when the mobile hydrocarbon-water interface is below the potentiometric surface or where only capillary hydrocarbon exists. Some of the major areas identified by Testa and Paczkowski (1989) where errors can also be introduced include:

- Accuracy of the measuring device used for the initial gauging and recovery of the levels after bailing
- Operator error in measuring and recording levels with time
- Inability of operator to collect early recovery data due to rapid rising well levels
- Bailing groundwater in addition to product from a low-yielding formation
- Lack of a theoretical response or inflection point due to an inordinate length of time for water recovery
- Variable accumulation rates of product caused by borehole effects
- Evaluation of type curves and selection of an inflection point
- Lack of cross-validation of methods

If bail-down testing has innate compounding errors within itself, these errors can only be further compounded since the remaining calculations, extrapola-

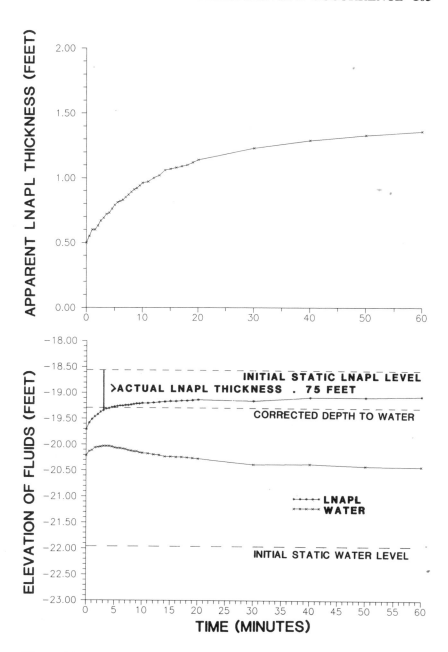

Figure 5.8. Representative bail-down testing curve results using Hughes et al. (1988) method where depth to top of product layer vs. time is produced.

tions, and evaluations are based upon this initial step. Although this procedure remains essentially unproven, it can be used as a useful supportive tool in comparing the actual thickness data generated from bail-down testing to those derived empirically, thus resulting in a range of values for the total mobile LNAPL product volume present.

When the elevation of the mobile LNAPL/water interface is not known, recharge testing has been employed to estimate the actual LNAPL thickness in the formation (Hughes et al., 1988). Similar to a conventional pumping test, mobile LNAPL is pumped from the well until steady-state conditions are reached, while maintaining the thinnest amount of hydrocarbon in the well as possible. No water is pumped from the well. The top of the liquid hydrocarbon surface is recorded as the hydrocarbon recharges. Elevation of the top of the fluid hydrocarbon surface vs. time is plotted. When the recharge rate begins to decline, this corresponds to the elevation of the entry point or inflection point. The actual thickness can then be calculated as discussed by Hughes et al., (1988) using the following equation:

$$Tf = \frac{Tc}{\left(1 - SGhc\right)}, \qquad (5.11)$$

where Tf = actual thickness of the mobile LNAPL in the formation; Tc = distance between the top of the mobile LNAPL layer and the buoyancy surface or inflection point; and SGhc = specific gravity of the LNAPL product.

The recharge test has the advantage of yielding results regardless of the location of the mobile LNAPL/water interface in relation to the potentiometric surface. Although more accurate than a bail-down test, it is also more complicated to conduct.

Innovative ways of determining actual LNAPL thickness in the formation are currently being explored and tested. In conjunction with bail-down testing, these techniques include dielectric well logging under unconfined conditions and optoelectronic sensing under confined conditions. Dielectric well logging allows detection of (1) interfaces between dissimilar fluids such as water, hydrocarbon, and air, (2) the relative dielectric constants of the fluids, and (3) well casing (nonperforated PVC casing) reflectance. Resolution is presently 5–6 in.; below this thickness a distinct hydrocarbon layer is not consistently discernable, although thicknesses as low as 3 in. have been resolved. Thus, this method is not well applicable to occurrences of limited relatively small thicknesses, but may prove promising for occurrences of large thicknesses.

Optoelectronic sensing is based on the ability to detect rising oil droplets that enter through the well screen openings and rise to the surface in a well. The device used incorporates the use of photosensors into solid-state electronic circuitry. To induce flow of mobile LNAPL into the well, measurement is conducted in conjunction with bailing of the well. The actual LNAPL product thickness (OT_F) is calculated by obtaining a•measurement of the apparent thickness in a well bore (OT_A) and then estimating the distance from the top of the product down to the confining layer (Ho). The relationship is as follows:

$$OT_F = (V_w) OT_A - Ho, \qquad (5.12)$$

where OT_F = actual LNAPL thickness; V_w = specific weight of the water gravity; OT_A = apparent LNAPL thickness; and Ho = depth of confining layer.

Potential problems can result if the monitoring well construction is such that the gravel pack and screened interval is not discretely terminated at the confining layer and grouted annulus thereof.

5.4   VOLUME DETERMINATION

Determination of actual LNAPL thickness in the formation is important in determining the true volume of product in the subsurface. This in turn allows for reasonable estimates of the time frame for recovery, as well as providing a mechanism for monitoring the efficiency and effectiveness of a recovery operation. The efficiency and effectiveness of many large-scale recovery operations are monitored by the reduction in volume with time. Thus, the percent reduction with time can easily be viewed as insignificant if exaggerated volumes are used. For example, if one estimates a volume on the order of 100,000 barrels, of which 10,000 barrels have been recovered to date, then a 10% reduction over a certain time interval has been achieved. However, if only 50,000 barrels exist, of which 10,000 barrels have been recovered, then a 20% reduction has actually been achieved.

Prior to the initiation of an LNAPL hydrocarbon recovery strategy, the total hydrocarbon and recoverable hydrocarbon volume is estimated. Volume estimates are dependent on the determination of actual mobile hydrocarbon thickness, which can be derived empirically and/or directly via field methods such as conducting bail-down or recharge tests. A simplistic hydrostatic model is provided by Testa and Paczkowski (1989). Initially, the measurement of apparent LNAPL product thicknesses as measured in monitoring wells is conducted. The data generated is then used to develop an "actual" hydrocarbon thickness contour (isopach) map. Once developed, planimetering is performed to derive the areal coverage of incremental apparent hydrocarbon thicknesses. The greater the coverage and number of data points (monitoring wells), the smaller the chosen increment for planimetering. Although apparent thicknesses can vary between monitoring points depending on the thickness of the capillary fringe, calculated thickness values between monitoring points are approximated. Thus, the capillary fringe, hence the apparent thickness, is assumed to be constant between monitoring points. Upon completion of planimetering, the volume of soil encompassed by the LNAPL product plume (Vs) is estimated. This value is then multiplied by an "assumed" porosity (ϕ) value, based on soil types encountered during the subsurface characterization process to grossly calculate the total apparent volume (Va) present as shown below:

$$Va = Vs \times \phi, \qquad (5.13)$$

where Va = total apparent volume of hydrocarbon present; Vs = volume of soil encompassed by LNAPL hydrocarbon product; and ϕ = porosity (assumed).

Equation 5.13 overpredicts the hydrocarbon volume, because in the volume of soil encompassed by LNAPL product, finite (and variable) water saturation exists, the minimum value being S_r (the irreducible water saturation). Hence, the maximum value of V_a based on V_s is

$$V_a(\max) = V_s \times \phi(1 - S_r) \text{ and} \qquad (5.13a)$$

$$V_a \qquad = V_s \times \frac{1}{z} \phi S_o. \qquad (5.13b)$$

In the capillary fringe, a third phase (air) can be present, further reducing the value of V_a.

Since the water table as measured in the well is depressed by the weight of the hydrocarbon, a corrected depth to water is calculated:

$$PTap = DTW - DTP \qquad (5.14)$$

$$CDTW = \text{Static DTW} - (PTap \times G), \qquad (5.15)$$

where CDTW = corrected depth to water; DTW = depth to water (measured); DTP = depth to product (measured); PTap = apparent product thickness (measured); and G = specific gravity of product at 60° F.

A correction factor is then applied for capillary fringe effects. This factor can be empirically derived, reflecting the corrected depth to water as shown below:

$$\text{Capillary fringe} (CF) = (CDTW - DTP) - PTac, \qquad (5.16)$$

where CF = capillary fringe thickness; CDTW = corrected depth to water (calculated); DTP = depth to product (measured); and PTac = actual product thickness (calculated).

The calculation of total apparent volume does not, however, take into consideration the specific yield of the formation. Specific yield is the percentage of the mobile free hydrocarbon product that will drain and be recovered under the influences of gravity. This value is dependent on flow characteristics of the hydrocarbon, as well as the characteristics of the formation. Typical values may range from 5 to 20%. The total apparent volume is multiplied by an "assumed" specific yield for the particular area to obtain the estimated volume of recoverable hydrocarbons:

$$H = Sy \times Va, \tag{5.17}$$

where H = recoverable hydrocarbon; Sy = specific yield; and Va = total apparent volume of hydrocarbon. More sophisticated methods for estimating LNAPL volumes have recently been proposed by Lenhard and Parker (1990), and Farr et al., (1990). The methods are based on the general approach that the specific volume of LNAPL can be determined by integrating the hydrocarbon saturation over the vertical distance of its occurrence:

$$V_o = \int_{Z_{ow}}^{Z_{ao}} \phi(z) S_o(z) dz, \tag{5.18}$$

where V_o is the specific volume (L^3/L^2); ϕ is the porosity (fraction); S_o is the saturation (fraction); z represents the vertical coordinate; and subscripts ao and ow denote the air/oil and oil/water interfaces, respectively. In turn, the saturation of LNAPL product phase is estimated from the capillary pressure-saturation correlations proposed by Brooks and Corey (1966) or van Genuchten (1980). The above relationships are in the following general form:

(a) Brooks-Corey:

$$S_o = f\left(P_c^{ao}, P_c^{ow}, \lambda, S_r\right). \tag{5.19}$$

(b) Van Genuchten:

$$S_o = f\left(\alpha_{ao}, \alpha_{ow}, {}^n, P_c^{ao}, P_c^{ow}, S_r\right). \tag{5.20}$$

In the above equations, P_c is the capillary pressure; λ is the Brooks-Corey pore-size distribution index; S_r is the residual saturation of the LNAPL product phase; α is the van Genuchten fluid/soil parameter; and the superscript n is the van Genuchten soil parameter.

Finally, the capillary pressure between phases is found by:

$$P_c^{ao} = f\left(\rho_o, D_w^{ao}, P_d^{ao}\right) \tag{5.21}$$

$$P_c^{ow} = f\left(\rho_o, \rho_w, D_w^{ow}, P_d^{ow}\right), \tag{5.22}$$

where D_w is the depth to the interface between two phases in the well; ρ is the phase density; and P_d is the displacement pressure. In equations 5.19 through 5.22, the superscripts a, o, and w denote air, oil (LNAPL product), and water phases.

In equation 5.18, S_o at Z_{ow} is maximum with the limiting value of $1-S_r$, where, S_r is the irreducible aqueous phase saturation in the porous medium under study. The value of S_o decreases with increasing elevation. Z_{ao}, the interface between air and the LNAPL product phase, may or may not coincide with Z_u, the upper boundary of the aquifer. Typically, the saturation of the LNAPL product phase extends over two distinct regions (see Figure 5.4). These are (1) water and free product phase zone and (2) water, free product phase, and air zone. When a single homogeneous stratum is considered, ϕ can be assumed constant. In a stratified medium, however, saturation discontinuities generally exist due to the variation in soil chracteristics, and the determination of LNAPL product volume based on equation 5.18 may become much more involved.

Although the approach is theoretically sound, both of the proposed relationships between capillary pressure and saturation equations 5.19 and 5.20 are highly nonlinear and are limited in practicality by the requirement of multiparameter identification. In addition, due to the inherent soil heterogeneities and difference in LNAPL composition, the identified parameters at one location cannot be automatically applied to another location at the same site, or less so at another site. For example, Farr et al., (1990) have reported the Brooks-Corey and van Genutchen parameters, λ, n, and α_{ao}, for seven different porous media based on least square regression of laboratory data. The parameters are found to vary about one order of magnitude and do not show any specific correlation for a particular soil type.

The general conclusions of the two studies are listed in the following:

1. The relationship between V_o, the LNAPL product volume, and H_o, the apparent thickness of LNAPL product in observation well is unique, nonlinear, and dependent on the particular porous medium/fluid system.
2. For relatively low values of apparent LNAPL product thickness (100 cm or less) observed in the well, the ratio of hydrocarbon volume to apparent product thickness is very low, in the range of 10^{-2} to 10^{-1} or even less.
3. For sufficiently large values of LNAPL product thickness, V_o/H_o approaches the limiting maximum value of $\phi(1-S_r)$, which is usually in the range of 0.2–0.4.
4. For an unconsolidated porous medium with uniform pore size distribution, the ratio of V_o/H_o approaches its maximum value much more rapidly than that found in a medium where pore size distribution is nonuniform, leading to the significant variation of water saturation in the contaminated zone.

An important assumption in the above approach is that the steady state (or equilibrium) condition exists among the three phases so that the laws of hydrostatics can be applied in the analysis. In an actual site, the process of attaining a true three-phase equilibrium is rather slow (and is not attained when there is periodic recharge and/or discharge), and a hydrodynamic model may be better suited for parameter identification and subsequent prediction. The limitations of hydrostatic models arise due to the fact that the height of the free product phase in the observation well is not directly related to the height of the free product in soil, except in the simplest case. When a well is being drilled at a contaminated site, factors affecting the radial flow of the LNAPL product phase into the well include the relative permeability of the immiscible phases as a function of phase saturation (and naturally, the elevation), absolute permeability of the porous medium, capillary pressure as a function of saturation, head difference, and wetability of the porous medium, among others. The LNAPL product phase flows into the well only from points in elevation (z axis) where the LNAPL saturation is sufficiently high, until an equilibrium in pressure in all directions within the observation well is attained.

REFERENCES

1. Acker, W. L., III, 1974, Basic Procedures for Soil Sampling and Core Drilling: ⬤Acker Drill Company, Scranton, PA, 246 p.
2. American Petroleum Institute (API), 1980, The Migration of Petroleum Products in the Soil and Groundwater. Principles and Countermeasures. API Publication No. 1628.
3. Ayra, L. M. and Paris, J. F., 1981, A Physico-Empirical Model to Predict Soil Moisture Characteristics from Particle-Size Distribution and Bulk Density Data: *Soil Sci. Soc. Am. J.*, Vol. 45, p. 1023-1030.
4. Bear, J., 1979, *Hydraulics of Groundwater*. McGraw-Hill, New York, New York, 569 p.
5. Blake, S. B. and Fryberger, J. S., 1983, Containment and Recovery of Refined Hydrocarbons from Groundwater: in Proceedings of Groundwater and Petroleum Hydrocarbons - Protection, Detection, Restoration. PACE, Toronto, Ontario.
6. Blake, S. B. and Hall, R. A., 1984, Monitoring Petroleum Spills with Wells: Some Problems and Solutions. In *Proceedings of the National Water Well Association of Ground Water Scientists and Engineers, 4th National Symposium on Aquifer Restoration and Groundwater Monitoring*, p. 305-310.
7. Brooks, R. H. and Corey, A. T., 1966, Properties of Porous Media Affecting Fluid Flow: Journal of Irrigation Drainage Division, ASCE, Vol. 92, No. IR2, p. 61-88.
8. Campbell, M. D., and Lehr, J. H. 1973, Water Well Technology, McGraw-Hill, New York, NY, 681 p.

9. Chafee, W. T. and Weimar, R. A., 1983, Remedial Programs for Ground-water Supplies Contaminated by Gasoline. In Proceedings of the National Water Well Association of Ground Water Scientists and Engineers, 3rd National Symposium on Aquifer Restoration and Groundwater Monitoring, p. 39-46.

10. CONCAWE, 1979, de Pastrovich, T. L., Baradat, Y., Barthel, R., Chiarelli, A., and Fussell, D. R., 1979, Protection of Groundwater from Oil Pollution. CONCAWE Report No. 3/79, The Hague, Netherlands, 61 p.

11. Corey, A. T., Rathjens, C. H., Henderson, J. H., and Wyllie, M. R. J., 1956, Three-Phase Relative Permeability: *Trans. Am. Insti. Mining Met. Petroleum Eng.*, Vol. 207, p. 349-351.

12. Corey, A. T., 1986, *Mechanics of Immiscible Fluids in Porous Media.* Water Resources Publications, Littleton, CO., 255 p.

13. Cummings, T. R. and Twenter, F. R., 1986, Assessment of Groundwater Contamination at Wurtsmith Air Force Base, Michigan, 1982-85: *U.S.G.S. Water Resources Investigations Report* 86-4188, 110 p.

14. Dietz, D. N., 1971, Pollution of Permeable Strata by Oil Components: In *Water Pollution by Oil* (edited by Hepple, P.), London, Institute of Petroleum, p. 127-139.

15. Driscoll, F. G., 1986, *Groundwater and Wells*: Johnson Division, St. Paul, MN, 2nd Edition, 1089 p.

16. Farr, A. M., Houghtalen, R. J., and McWhorter, D. B., 1990, Volume Estimation of Light Nonaqueous Phase Liquids in Porous Media: *Ground Water,* Vol. 28, No. 1, p. 48-56.

17. Faust, C. R., Guswa, J. H., and Mercer, J. W., 1989, Simulation of Three-Dimensional Flow of Immiscible Fluids Within and Below the Unsaturated Zone: *Water Resour. Res.,* Vol. 25, No. 12, p. 2449-2464.

18. Folkes, D. J., Bergman, M. S., and Herst, W. E., 1987, Detection and Delineation of a Fuel Oil Plume in a Layered Bedrock Deposit: In Proceedings of the National Water Well Association of Ground Water Scientists and Engineers and the American Petroleum Institute Conference on Petroleum Hydrocarbons and Organic Chemicals in Ground Water: Prevention, Detection and Restoration, November, 1987, p. 279-304.

19. Gruszczenski, T. S., 1987, Determination of a Realistic Estimate of the Actual Formation Product Thickness Using Monitor Wells: A Field Bailout Test. In Proceedings of the National Water Well Association of Ground Water Scientists and Engineers and the American Petroleum Institute Conference on Petroleum Hydrocarbons and Organic Chemicals in Ground Water: Prevention, Detection and Restoration, November, 1987, p. 235-253.

20. Hall, R., Blake, S. B., and Champlin, S. C., Jr., 1984, Determination of Hydrocarbon Thickness in Sediments Using Borehole Data: In Proceedings of the National Water Well Association of Ground Water Scientists and Engineers, Fourth National Symposium on Aquifer Restoration and Groundwater Monitoring, p. 300-304.

21. Hampton, D. R., 1988, Laboratory and Field Comparisons between Actual and Apparent Product Thickness in Sands: American Geophysical Union Abstract, Fall Meeting, December, 1988, EOS, Vol. 69, No. 44, p. 1213.

22. Hampton, D. R., 1989, Laboratory Investigation of the Relationship Between Actual and Apparent Product Thickness in Sands. In Proceedings of Symposium Conference on Environmental Concerns in the Petroleum Industry (Edited by Testa, S.M.), Pacific Section American Association of Petroleum Geologists, p. 31-55.

23. Hampton, D. R. and Miller, P. D. R., 1988, Laboratory Investigations of the Relationship Between Actual and Apparent Product Thickness in Sands: In Proceedings of the National Water Well Association of Ground Water Scientists and Engineers and the American Petroleum Institute Conference on Petroleum Hydrocarbons and Organic Chemicals in Ground Water: Prevention, Detection and Restoration, Vol. I, November, 1988, p. 157-181.

24. Hampton, D. R., Wagner, R. B., and Heuvelhorst, H. G., 1990, A New Tool to Measure Petroleum Thickness in Shallow Aquifers: In Proceedings of the National Water Well Association of Groundwater Scientists and Engineers Fourth National Outdoor Action Conference on Aquifer Restoration, Ground Water Monitoring and Geophysical Methods, May 1990, in press.

25. Hughes, J. P., Sullivan, C. R., and Zinner, R. E., 1988, Two Techniques for Determining the True Hydrocarbon Thickness in an Unconfined Sandy Aquifer: In Proceedings of the National Water Well Association of Ground Water Scientists and Engineers and the American Petroleum Institute Conference on Petroleum Hydrocarbons and Organic Chemicals in Ground Water: Prevention, Detection and Restoration, Vol. I, November, 1988, p. 291-314.

26. Hunt, W. T., Wiegand, J. W., and Trompeter, J. D., 1989, Free Gasoline Thickness in Monitoring Wells Related to Ground Water Elevation Change: In Proceedings of the Auburn University Water Resources Research Institute Conference on New Field Techniques for Quantifying the Physical and Chemical Properties of Heterogeneous Aquifers, National Water Well Association, March, 1989, p. 671-692.

27. Keech, D. H., 1988, Hydrocarbon Thickness on Groundwater by Dielectric Well Logging: In Proceedings of the National Water Well Association of Ground Water Scientists and Engineers and the American Petroleum Institute Conference on Petroleum Hydrocarbons and Organic Chemicals in Ground Water: Prevention, Detection and Restoration, Vol. I, November, 1988, p. 275-289.

28. Kemblowski, M. W. and Chiang, C. Y., 1990, Hydrocarbon Thickness Fluctuations in Monitoring Wells: *Ground Water*, Vol. 28, No. 2, p. 244-252.

29. Kemblowski, M. W. and Chiang, C. Y., 1988, Analysis of the Measured Free Product Thickness in Dynamic Aquifers: In Proceedings of the National Water Well Association of Ground Water Scientists and Engineers and the American Petroleum Institute Conference on Petroleum Hydrocarbons and Organic Chemicals in Ground Water: Prevention, Detection and Restoration, Vol. I, November, 1988, p. 183-205.

30. Kessler, A. and Rubin, H., 1987, Relationships Between Water Infiltration and Oil Spill Migration in Sandy Soils. *J. Hydrol.*, Vol. 91, p. 187-204.

31. Kimberlin, D. K. and Trimmell, M. L., 1988, Utilization of Optoelectronic Sensing to Determine Hydrocarbon Thicknesses within Confined Aquifers: In Proceedings of the National Water Well Association of Ground Water Scientists and Engineers and the American Petroleum Institute Conference on Petroleum Hydrocarbons and Organic Chemicals in Ground Water: Prevention, Detection and Restoration, Vol. I, November, 1988, p. 255-274.

32. Kool, J. B., Parker, J. C., 1987, Development and Evaluation of Closed-Form Expressions for Hysteretic Soil Hydraulic Properties: *Water Resourc. Res.*, Vol. 24, p. 105-114.

33. Kramer, W. H., 1982, Groundwater Pollution From Gasoline. *Groundwater Monitoring Review*, Vol. 2(2), p. 18-22.

34. Laliberte, G. E., Corey, A. T., and Brooks, R. H., 1966, Properties of Unsaturated Porous Media: Colorado State University, Fort Collins, Hydrology Papers No. 17.

35. Lenhard, R. J. and Parker, J. C., 1987, Measurement and Prediction of Saturation-Pressure Relationships in Three-Phase Porous Media Systems: *Journal of Contaminant Hydrology*, Vol. 1, p. 407-424.

36. Lenhard, R. J. and Parker, J. C., 1988, Experimental Validation of the Theory of Extending Two-Phase Saturation-Pressure Relationships to Three-Fluid Phase System for Monotonic Drainage Paths: *Water Resour. Res.*, Vol. 24, p. 373-380.

37. Lenhard, R. J., Dane, J. H., Parker, J. C., and Kaluarachchi, 1988a, Measurement and Simulation of One-Dimensional Transient Three-Phase Flow for Monotonic Liquid Drainage: *Water Resour. Res.*, Vol. 24, p. 853-863.

38. Lenhard, R. J. and Parker, J. C., 1990, Estimation of Free Hydrocarbon Volume from Fluid Levels in Observation Wells: *Ground Water*, Vol. 28, No. 1, p. 57-67.
39. Littlefield, K. V., Wehler, N. E., and Heard, R. W., 1984, Identification and Removal of Hydrocarbons from Unconsolidated Sediments Affected by Tidal Fluctuations: In Proceedings of the National Water Well Association of Ground Water Scientists and Engineers, Fourth National Symposium on Aquifer Restoration and Groundwater Monitoring, p. 316-322.
40. Mansure, C. and Fouse, J., 1984, Design of Hydrocarbon Recovery System, Miami International Airport, In Proceedings of the National Water Well Association and American Petroleum Institute Conference on Petroleum Hydrocarbons and Organic Chemicals in Groundwater: Prevention, Detection and Restoration, p. 400-419.
41. Mull, R., 1978, Calculations and Experimental Investigations of the Migration of Oil Products in Natural Soils: In International Symposium on Ground Water Pollution by Oil Hydrocarbons, Proceedings of International Association of Hydrogeologists, Prague, June 5-7, 1978, p. 167-181.
42. Parker, J. C., Lenhard, R. J., and Kuppusamy, T., 1987, A Parametric Model for Constitutive Properties Governing Multiphase Flow in Porous Media. *Water Resour. Res.*, Vol. 23(4), p. 618-624.
43. Parker, J. C. and Lenhard, R. J., 1989, Vertical Integration of Three Phase Flow Equations for Analysis of Light Hydrocarbon Plume Movement: Transport in Porous Media (in review).
44. Schiegg, H. O., 1985, Considerations on Water, Oil, and Air in Porous Media: *Water Science and Technology*, Vol. 17 (4-5), p. 467-476.
45. Schwille, F., 1967, Petroleum Contamination of the Subsoil—a Hydrological Problem: In P. Hepple, Ed., The Joint Problems of the Oil and Water Industries: London Institute of Petroleum, p. 23-54.
46. Shephard, W. D., 1983, Practical Geohydrological Aspects of Ground-Water Contamination: In Proceedings of the National Water Well Association of Ground Water Scientists and Engineers, Third National Symposium on Aquifer Restoration and Ground Water Monitoring, p. 365-372.
47. Testa, S. M. and Paczkowski, M. T., 1989, Volume Determination and Recoverability of Free Hydrocarbon: *Groundwater Monitoring Review*, Winter Issue, p. 120-128.
48. Trimmell, M. L., 1987, Installation of Hydrocarbon Detection Wells and Volumetric Calculations Within a Confined Aquifer: In Proceedings of the National Water Well Association of Ground Water Scientists and Engineers and the American Petroleum Institute Conference on Petroleum Hydrocarbons and Organic Chemicals in Ground Water: Prevention, Detection and Restoration, November, 1987, p. 255-269.

49. van Dam, 1967, The Migration of Hydrocarbons in a Water-Bearing Stratum: In P. Hepple, ed., The Joint Problems of the Oil and Water Industries: London, Institute of Petroleum, p. 55-96.

50. van Genuchten, M. Th., 1980, A Closed-Form Equation for Predicting the Hydraulic Conductivity of Unsaturated Soils: Soil Sci. Soc. Am. J., Vol. 44, p. 892-898.

51. Wagner, R. B., Hampton, D. R., and Howell, J. A., 1989, A New Tool to Determine the Actual Thickness of Free Product in a Shallow Aquifer: In Proceedings of the National Water Well Association of Ground Water Scientists and Engineers and the American Petroleum Institute Conference on Petroleum Hydrocarbons and Organic Chemicals in Ground Water: Prevention, Detection and Restoration, November, 1989, p. 45-59.

52. White, N. F., 1968, The Desaturation of Porous Materials: Ph.D. dissertation, Colorado State University, Fort Collins.

53. Williams, D. E. and Wilder, D. G., 1971, Gasoline Pollution of a Groundwater Reservoir - A Case History: Ground Water, Vol. 9, No. 6, p. 50-56.

54. Wilson, J. L., Conrad, S. H., Hagan, E., Mason, W. R., and Peplinski, W., 1988, The Pore Level Spatial Distribution and Saturation of Organic Liquids in Porous Media: In Proceedings of the National Water Well Association of Ground Water Scientists and Engineers and the American Petroleum Institute Conference on Petroleum Hydrocarbons and Organic Chemicals in Ground Water: Prevention, Detection and Restoration, Vol. 1, November, 1988, p. 107-133.

55. Yaniga, P. M., 1982, Alternatives in Decontamination for Hydrocarbon - Contaminated Aquifers: Ground Water Monitoring Review, Vol. 2, p. 40-49.

56. Yaniga, P. M., 1984, Hydrocarbon Retrieval and Apparent Hydrocarbon Thickness: Interrelationships to Recharging/Discharging Aquifer Conditions: In Proceedings of the National Water Well Association of Ground Water Scientists and Engineers and the American Petroleum Institute Conference on Petroleum Hydrocarbons and Organic Chemicals in Ground Water: Prevention, Detection and Restoration, November, 1984, p. 299-329.

57. Yaniga, P. M. and Demko, D. J., 1983, Hydrocarbon Contamination of Carbonate Aquifers: Assessment and Abatement: In Proceedings of the National Water Well Association of Ground Water Scientists and Engineers Third National Symposium on Aquifer Restoration, p. 60-65.

58. Zilliox, L. and Muntzer, P., 1975, Effects of Hydrodynamic Processes on the Development of Ground-Water Pollution: Study on Physical Models in a Saturated Porous Medium. Progress in Water Technology, vol. 7, p. 561-568.

6 RECOVERABILITY OF LNAPL HYDROCARBON PRODUCT

"Very little of what hydrocarbon is in the ground is recoverable via conventional techniques"

6.1 INTRODUCTION

Remediation of petroleum contaminated aquifers where LNAPL product is present is typically a two-phase process. The initial phase consists of removal of as much LNAPL product as possible. The next phase is relatively more difficult, consisting of secondary displacement of residual hydrocarbons and subsequent removal of dissolved and gaseous components. The time frame associated with the second phase is dependent on the volume of residual hydrocarbon that remains following initial recovery and is further discussed in later chapters. Primary and secondary factors associated with the recoverability of LNAPL are outlined in Table 6.1 and are discussed below.

6.2 RELATIVE PERMEABILITY

The potential for recovery of LNAPL product is governed by several factors, including viscosity, density, actual saturated thickness of the hydrocarbon in the formation, residual water saturation, and the permeability of the formation. These factors determine the relative permeability of the formation to the hydrocarbon. The relative permeability is a measure of the relative ability of hydrocarbon and water to migrate through the formation as compared to a single

Table 6.1   Factors Associated with Recovery of LNAPL

Primary Factors	Secondary Factors
Relative permeability (based on percent water saturation)	Amount of residual hydrocarbon
Product viscosity	Areal distribution of LNAPL pool
	Site-specific constraints

fluid. It is expressed as a fraction or percentage of the permeability in a single fluid system. Relative permeability can be determined experimentally for each formation material and for each combination of fluid saturations and fluid properties. During hydrocarbon recovery, their ratios are constantly changing.

Graphs of relative permeability are generally similar in pattern to that shown in Figure 6.1. As shown, some residual water remains in the pore spaces, but water does not begin to flow until its water saturation reaches 20% or greater. Water at the low saturation is interstitial or "pore" water which preferentially wets the material and fills the finer pores. As water saturation increases from 5 to 20%, hydrocarbon saturation decreases from 95 to 80% where, to this point, the formation permits only hydrocarbon to flow, not water. Where the curves cross (at a saturation of 56% for water and 44% for hydrocarbon) the relative permeability is the same for both fluids. Both fluids flow, but at a level of less than 30% of what each fluid flow would be at 100% saturation. With increasing water saturation, water flows more freely and hydrocarbon flow decreases. When the hydrocarbon saturation approaches 10%, the hydrocarbon becomes immobile, allowing only water to flow. For the example given, the hydrocarbon residual saturation is 10% pore saturation limited by the fluid density and viscosity and the formation permeability.

The relations shown in Figure 6.1 have a wide application to problems of fluid flow through permeable material. One of the most important applications for recovery of hydrocarbon is that there must be at least 5–10% saturation with the nonwetting fluid, and 20–40% saturation with the wetting fluid before flow occurs. Thus for hydrocarbon (the nonwetting fluid), there must be a minimum of 5–10% saturation of the pore space before the fluid can move through the partially saturated or unsaturated formation and accumulate. Conversely, every hydrocarbon accumulation has a quantity of hydrocarbon that is not mobile, since it is at or below a saturation of 5–10% and is thus not recoverable.

The mixed flow of water and LNAPL to a recovery well at a restoration site is very similar to oil and water flow to a pumping well in a low-pressure water-drive reservoir. The equation used by Muskat (1937) to describe flow to a production well is presented in the following equation:

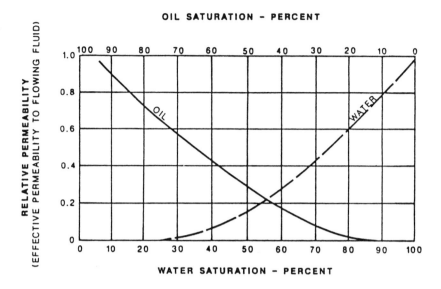

Figure 6.1.   Relative permeability curves for oil and water (after Levorsen, 1967).

$$Q = \frac{2\pi k T P,}{\mu \ln\left(\dfrac{L}{r}\right)},$$ (6.1)

where Q = rate of flow and k = intrinsic permeability:

$$N.B. \quad k = \frac{K\mu}{\checkmark}$$

where $\checkmark$ = specific weight; μ = viscosity; K = normal permeability; L = radius over which drainage occurs; T = thickness of reservoir; P = pressure drop over distance L; μ = viscosity; and r = well bore radius.

Note the similarity to the Dupuit equation for confined aquifers:

$$Q = 2\,\pi K T \frac{h_o - h_w}{\ln\left(\dfrac{L}{r}\right)}$$ (6.2)

where K = normal permeability and $h_o - h_w$ = pressure drop (in water head) between L and r.

When the Muskat and Dupuit equations are combined and then integrated to approximate steady-state flow to an unconfined aquifer (as found at most remedial sites), the resulting equation takes the following form:

$$Q = \frac{\pi k \left(h_o^2 - h_w^2 \right)}{\mu \ln \left(\dfrac{L}{r} \right)} \tag{6.3}$$

where k = intrinsic permeability; μ = viscosity; h_o = water head equivalent at distance L; and h_w = water head equivalent at well. N.B. correction for fluid density is included in the $h_o - h_w$ factor.

If the values for all the factors are known and expressed in common units, this equation will provide a reasonable estimate of flow from recovery wells. The difficulty in the actual application of this equation is the determination of the water/oil saturation in the aquifer matrix (thus, the relative permeability). The equation assumes a steady state flow setting. In a dynamic situation where the oil reserves are being depleted and water/oil mixtures are variable, it is almost impossible (from an economic point of view) to use these equations precisely. However, inclusion of some assumptions based on results of the initial site investigation can often be of assistance in the initial spacing of recovery wells and estimating recovery rates.

It should be recognized that observations made in monitoring wells surrounding the recovery wells must be adjusted to reflect the true water head pressure due to varying LNAPL thickness in these wells over time. With recovery, the apparent product thickness may increase (or decrease) as the true formation thickness decreases.

6.3 RESIDUAL HYDROCARBON

The recoverability of hydrocarbon from the subsurface refers to the amount of mobile hydrocarbon available. Hydrocarbon that is retained in the unsaturated zone is not typically recoverable by conventional means. Additional amounts of hydrocarbon that are unrecoverable by conventional methods include the immobile hydrocarbons associated with the water table capillary zone. Residual hydrocarbon is pellicular or insular, and is retained in the aquifer matrix. In respect to recoverability, residual hydrocarbons entrapment can result in volume estimate discrepancies, as well as a decrease in recovery efficiency. With increasing water saturation, such as when the water table rises via recharge or product removal, hydrocarbons essentially become occluded by a continuous water phase. This results in a reduction of LNAPL and product thickness as measured in the well at constant volume. When water saturation is decreased by lowering the water table (as during recovery operations), trapped hydrocarbons can remobilize, leading to increased recoverability.

In general, as viscosity of the hydrocarbon increases and grain size decreases, the residual saturation increases. Typical residual saturation values for unsaturated, porous soil are presented by CONCAWE (1979) and are tabulated in Table 4.2.

These values are then multiplied by a correction factor to account for hydrocarbon viscosity. Correction factors for different hydrocarbon types (CONCAWE, 1979) are

- 0.5 for low viscosity products (gasoline)
- 1.0 for kerosene and gas oil
- 2.0 for more viscous oils

The American Petroleum Institute (1980) has presented some similar guidelines for estimating residual saturation. Basing their work on a "typical" soil with a porosity of 30% the API gives residual saturation values noted as a percentage of the total porosity of the soil as follows:

- 0.18 for light oil and gasoline
- 0.15 for diesel and light fuel oil
- 0.20 for lube and heavy fuel oils

Similar studies done by Hall et al., (1984) on hydrocarbon of lower API gravities (i.e., gravities between 34° and 40°) show that specific retention for more viscous hydrocarbons can range between 35% and 50% of the pore volume for fine sands with porosities of approximately 30%. The loss due to retention in the aquifer as the hydrocarbon migrates to the recovery well can be significant. Wilson and Conrad (1984) claim that residual losses are much higher in the saturated zone (i.e., capillary zone) than in the unsaturated zone.

Comparisons of the estimated volume to the actual volume recovered proves to be the only reasonable procedure for assessing the recoverable volume, considering all the variables involved. These comparisons indicate that the volume of hydrocarbon retained in the aquifer is higher than published residual saturation values. Based on the experience with gasoline and low viscosity hydrocarbons, the recoverable volumes have ranged from 20% to 60% of the pore volume in fine to medium sands.

6.4   OTHER FACTORS

In addition to factors concerning relative permeability, viscosity, and residual hydrocarbon, areal distribution of the plume and site-specific physical constraints can have a significant impact upon the degree of recoverability. A relatively small plume in areal extent with concentrated thicknesses is more recoverable, for example, than a thin plume with a large areal distribution. Site specific physical constraints may have a major impact upon the recoverability of the plume. The problem centers around the difficulty in locating recovery well(s) in their optimum location without conflicting with the facility layout. Furthermore, most recovery programs generate contaminated groundwater. Depending on the size of the facility and the scale of the recovery project, the recoverability of hydrocarbon and the respective time frame may be limited and may be highly dependent on the amount of water the facility can handle, and the subsequent treatment and disposal options available.

6.5   TIME FRAME FOR RECOVERY

The length of time required for recovery of LNAPL products is based on the volume of product present in the subsurface and is limited by numerous factors. Most often based on an educated guess, factors regarding volume estimates have innate compounding errors in relation to

- Accuracy of physical measurement where high viscosity and emulsified hydrocarbon is encountered
- Determination of true vs. apparent thickness
- Validity of bail-down tests for estimation of true thickness
- Extrapolation of geologic and hydrogeologic information between moni toring points
- Averaging of apparent thickness for planimetering
- Estimation or assumption for key factors including porosity, specific yield and retention values
- Variable product density
- Effects of residual trapped hydrocarbons

After the initial volume estimate has been determined, testing of a pilot recovery system should be initiated to evaluate recovery rates. However, factors that significantly affect recovery rates include the areal distribution and geometry of the free hydrocarbon product plume, type(s) and design of recovery system selected, and the performance and efficiency of the system with time.

Volume determinations and the subsequent time frame for recovery of LNAPL product can be estimated. However, a large number of compounding errors are associated with these calculations, thus a reasonable time frame for remediation is clearly an estimate.

The progress of recovery efforts cannot be based confidently on LNAPL product thickness maps. Although these maps provide a quantification of overall trends, the numerous factors that impact hydrocarbon thicknesses make accurate quantification difficult. An estimate of effectiveness thus is based on volume recovered to date, divided by the total volume that is considered recoverable. Furthermore, as the recovery project progresses and new data are introduced, the volume and time frame for recovery should be continually reevaluated and revised.

A method of estimating the volume of recoverable product after recovery is contouring the LNAPL thickness in monitoring wells and calculating the apparently oil-saturated soil volume at two separate times. The difference in the apparent oil-saturated soil volumes at these times is attributable to the quantity of oil recovered during that period. Using that relationship, an extrapolation can be used to estimate the remaining volume of recoverable oil. This value, in most situations, results in an overstatement of the recoverable volume of LNAPL in place.

In determining the total and recoverable volumes of LNAPL, the factor of recharge to the volume is undeterminable, but realistic. Developing a range of total and recoverable volumes is suggested. A valid way to determine this range is a comparison of values generated from the indirect empirical and direct field methods. Also, as additional monitoring well points are incorporated into the project, these new data are coupled with existing data and revised estimates are made. Finally, comparisons of the estimated recoverable volumes to the actual volume produced proves to be the only reasonable procedure for estimating the recoverable volume, considering all the variables involved.

REFERENCES

1. American Petroleum Institute (API), 1980, The Migration of Petroleum Products in the Soil and Groundwater. Principles and Countermeasures: *API Publication*, No. 1628.
2. Charbeneau, R. J., Wanakule, N., Chiang, C. Y., Nevin, J. P., and Klein, C. L., 1989, A Two-Layer Model to Simulate Floating Free Product Recovery: Formulation and Applications: In Proceedings of the National Water Well Association and American Petroleum Institute Conference on Petroleum Hydrocarbons and Organic Chemicals in Ground Water: Prevention, Detection and Restoration, November, 1989, p. 333-345.
3. de Pastrovich, T. L., Baradat, Y., Barthel, R., Chiarelli, A., and Fussell, D. R., (CONCAWE), 1979, Protection of Groundwater from Oil Pollution. CONCAWE Report No. 3/79, The Hague, Netherlands, 61 p.

4. Gruszczenski, T. S., 1987, Determination of a Realistic Estimate of the Actual Formation Product Thickness Using Monitor Wells: A Field Bailout Test. In Proceedings of the National Water Well Association of Ground Water Scientists and Engineers and the American Petroleum Institute Conference on Petroleum Hydrocarbons and Organic Chemicals in Ground Water: Prevention, Detection and Restoration, November, 1987, p. 235-253.

5. Hall, R., Blake, S. B., and Champlin, S. C., Jr., 1984, Determination of Hydrocarbon Thickness in Sediments Using Borehole Data: In Proceedings of the National Water Well Association of Ground Water Scientists and Engineers, Fourth National Symposium on Aquifer Restoration and Groundwater Monitoring, p. 300-304.

6. Hampton, D. R., 1989, Laboratory Investigation of the Relationship Between Actual and Apparent Product Thickness in Sands. In Proceedings of Symposium Conference on Environmental Concerns in the Petroleum Industry (Edited by Testa, S.M.), Pacific Section American Association of Petroleum Geologists, p. 31-55.

7. Kaluarachchi, J. J. and Parker, J. C., 1989, An Efficient Finite Element Method for Modeling Multiphase Flow: *Water Resour. Res.*, Vol. 25, p. 43-54.

8. Kaluarachchi, J. J., Parker, J. C. and Lenhard, R. J., 1989, A Numerical Model for Areal Migration of Water and Light Hydrocarbon in Unconfined Aquifers: *Advanced Water Resources,* in press.

9. Kool, J. B., Parker, J. C. and van Genuchten, M. Th., 1987, Parameter Estimation for Unsaturated Flow and Transport Models: A Review, Journal of Hydrology, Vol. 91, p. 255-293.

10. Kool, J. B. and Parker, J. C., 1988, Analysis of the Inverse Problem for Unsaturated Transient Flow, *Water Resour. Res.,* Vol. 24, p. 814-830.

11. Lenhard, R. J. and Parker, J. C., 1987a, A Model for Hysteretic Constitutive Relations Governing Multiphase Flow, 2. Permeability-Saturation Relations: *Water Resour. Res.,* Vol. 23, p. 2197-2206.

12. Lenhard, R. J. and Parker, J. C., 1987b, Measurement and Prediction of Saturation-Pressure Relationships in Three-Phase Porous Media Systems: *Journal of Contaminate Hydrology,* 1, p. 407-424.

13. Lenhard, R. J. and Parker, J. C., 1988, Experimental Validation of the Theory of Extending Two-Phase Saturation-Pressure Relations to Three-Fluid Phase Systems for Monotonic Drainage Paths: *Water Resour. Res.,* Vol. 24, p. 373-380.

14. Lenhard, R. J., Parker, J. C., and Kaluarachchi, J. J., 1989, A Model for Hysteretic Constitutive Relations Governing Multiphase Flow, 3. Refinements and Numerical Simulations: *Water Resour. Res.,* Vol. 25, p. 1727-1736.

15. Lenhard, R. J. and Parker, J. C., 1989, Estimation of Free Hydrocarbon Volume from Fluid Levels in Observation Wells: *Ground Water*, in press.
16. Leverett, M. C., 1939, Flow of Oil-Water Mixtures Through Unconsolidated Sands: American Institute of Mining and Metallurgy Engineering, Petroleum Development Technology, Vol. 132, p. 149-171.
17. Levorsen, A. I., 1967, *Geology of Petroleum:* W. H. Freeman and Company, New York, New York, 724 p.
18. Littlefield, K. V., Wehler, N. E., and Heard, R. W., 1984, Dentification and Removal of Hydrocarbons from Unconsolidated Sediments affected by Tidal Fluctuations: In Proceedings of the National Water Well Association Fourth National Symposium on Aquifer Restoration and Ground Water Monitoring, p. 316-322.
19. Mishra, S., Parker, J. C., and Kaluarachchi, J. J., 1989, Analysis of Uncertainty in Predictions of Hydrocarbon Recovery from Spill Sites: *Journal of Contaminate Hydrology*, in review.
20. Mishra, S., and Parker, J. C., 1989, Effects of Parameter Uncertainty on Predictions of Unsaturated Flow: *Journal of Hydrology,* Vol. 108, p. 19-33.
21. Mishra, S., Parker, J. C. and Singhal, N., 1989, Estimation of Soil Hydraulic Properties and Their Uncertainty from Particle Size Distribution Data: *Journal of Hydrology,* Vol. 108, p. 1-18.
22. Muskat, M., 1937, *Flow of Homogeneous Fluids Through Porous Media:* McGraw Hill, New York, 763 p.
23. Parker, J. C. and Lenhard, R. J., 1987, A Model for Hysteretic Constitutive Relations Governing Multiphase Flow, 1. Saturation-Pressure Relations, *Water Resour. Res.,* Vol. 23, p. 2187-2196.
24. Parker, J. C., Kaluarachchi, J. J. and Katyal, A. K., 1988, Areal Simulations of Free Product Recovery from a Gasoline Storage Tank Leak: In Proceedings of the National Water Well Conference on Petroleum Hydrocarbons and Organic Chemicals in Groundwater, November, 1988, p. 315-332.
25. Parker, J. C. and Kaluarachchi, J. J., 1989, A Numerical Model for Design of Free Product Recovery Systems at Hydrocarbon Spill Sites: In Proceedings of the National Water Well Association Fourth International Conference on Solving Groundwater Problems with Models.
26. Parker, J. C. and Lenhard, R. J., 1989, Vertical Integration of Three Phase Flow Equations for Analysis of Light Hydrocarbon Plume Movement: *Transport in Porous Media,* in review.
27. Schiegg, H. O., 1985, Considerations on Water, Oil, and Air in Porous Media: *Water Science and Technology,* Vol. 17 (4-5), p. 467-476.
28. Testa, S. M. and Paczkowski, M. T., 1989, Volume Determination and Recoverability of Free Hydrocarbon: *Ground Water Monitoring Review,* Winter Issue, Vol. 9, No. 1, p. 120-128.

29. Todd, D. K., 1959, *Ground Water Hydrology:* John Wiley, New York, 336 p.

30. van Dam, J., 1967, The Migration of Hydrocarbons in a Water-Bearing Stratum: In P. Hepple, Ed., *The Joint Problems of the Oil and Water Industries:* London, Institute of Petroleum, p. 55-96.

31. Yaniga, P. M. and Warburton, J. G., 1984, Discrimination Between Real and Apparent Accumulation of Immiscible Hydrocarbons on the Water Table: A Theoretical and Empirical Analysis, In Proceedings of the National Water Well Association Fourth National Symposium and Exposition on Aquifer Restoration and Ground Water Monitoring, p. 311-315.

7 REMEDIAL TECHNOLOGIES FOR RECOVERY OF LNAPL HYDROCARBON

"Given a hydrogeologic setting, a number of general and site specific considerations must be made in devising a rational and cost-effective aquifer restoration program tailored to its unique needs"

Recovery of spilled hydrocarbons has been occurring almost as long as petroleum has been refined. The earliest attempt reported was the use of pitcher pumps attached to shallow post-hole depth wells along a leaking pipeline. This pre-1900 effort was not driven by environmental concerns, but by the profit motive. Most recovery efforts were continued until the labor value exceeded the product value and then stopped. Primitive equipment, coupled with a lack of understanding of the mechanics of petroleum migration in freshwater aquifers and the relatively low value of the recovered product hampered the development of remedial technology.

Oil-field production technology continued to develop at a steady pace, since the large volumes of oil produced justified the expense of research and development. Improvements in spilled petroleum recovery technology were implemented when a few very large spills were discovered, since recovery proved economically rewarding. Subsequent environmental regulations, as discussed in Chapter 2, provided the mechanism requiring recovery of leaked product. Since these stimuli were introduced, many ingenious procedures have been implemented for recovery at specific site locations. Most of the equipment designed for these purposes were based on existing oil field, refinery, or water well technology (although modified in scale and complexity).

This chapter presents some of the more conventional LNAPL remediation equipment, along with a general discussion of the applicability of each. As in all design decisions, any one procedure or piece of equipment may not be totally adequate to render a solution to the problem. The designer is encouraged to mix and match to suit the particular needs of the site at hand. With new equipment being introduced at a very rapid rate, manufacturers catalogs are a significant aid to keeping abreast of currently available hardware and technology. A summary of more conventional approaches to the recovery of LNAPL is presented in Table 7.1 and outlined and discussed below:

- Linear interception (passive and active) systems
  - passive
  - active
- Well-point systems
- Vacuum-enhanced suction-lift well-point systems
- One-pump systems
  - submersible turbine (electrical)
  - mechanical lift
  - positive displacement pumps
- Two-pump systems
- Skimming units
  - floating skimmers
  - suspended skimmers
- Others
  - timed bailers
  - rope skimmers
  - belt skimmers
  - vacuum-assisted
  - vapor extraction
  - biodegradation

7.1   LINEAR INTERCEPTION

At favorable locations, the use of linear interception procedures has been effective to recover both LNAPL and dissolved contaminants. The concept behind ditches, trenches, and hydraulic troughs is to create a long area of low hydraulic head, which causes subsurface flow to be directed to a recovery location. Line sinks, such as engineered trenches, drains, and ditches, are the subject of this section. A hydraulic trough can also be created by installing a linear set of wells with overlapping cones of depression, as discussed in a later section.

Site conditions favorable for the construction of both trenches and ditches are generally those that are suitable for agricultural drainage. Most functioning line sink systems are based on the principles of drainage design. Typically, excavated linear systems are constructed where

- The area has unrestricted access by excavating equipment
- Air quality concerns from vapors are minimal
- Water/product depth is less than 25 ft.
- Few or no subsurface utilities or excavation restrictions are present

Interceptor trenches are generally constructed on the down-gradient side of a contaminant plume/pool perpendicular to the flow. As groundwater flows into or across the interceptor trench, LNAPL product is exposed for collection and contaminated water is available for collection and treatment. A wide variety of interceptor designs have been used, including both passive (i.e., no external flow stimulation) and active (i.e., induced flow such as pumped outlets).

7.1.1   Passive Systems

Almost all passive systems are open trenches that rely on the natural hydraulic gradient to transport LNAPL product for recovery. Under normal circumstances, the flow of LNAPL product into this type of system is very slow. Low permeability soils often do not conduct fluids rapidly enough to collect free volatile product, because the product evaporates from the exposed surface before it accumulates sufficiently to be collected. Highly permeable soils typically are subject to a low hydraulic gradient, which limits the rate of flow into the trench. Conditions that are more favorable to passive trench recovery are schematically shown in Figure 7.1 and include

- Moderate to high permeability soils
- Favorable hydraulic gradient
- Small seasonal water table fluctuation

A passive trench system is often considered suitable as an interim procedure prior to active remediation. Its advantages include the relative ease, cost of construction (using conventional construction equipment), and low maintenance effort.

Major disadvantages include the slow rate of recovery, exposure of large surface areas of flammable fluids, minimal containment, and the odor or air quality concerns. Additionally, this recovery method does not address the treatment of dissolved hydrocarbon constituents in groundwater.

Table 7.1   Summary of Conventional LNAPL Hydrocarbon Recovery Approaches

Aquifer Restoration Approach	Type	Hydrogeologic Conditions[a] (permeability)	Depth to Water (feet)	LNAPL	Dissolved Hydrocarbons	Restrictions
Linear interception	Passive		0–10	Yes	No	Loosely consolidated formations; underground structures
	Active		0–10	Yes	Yes	
One-pump system	Submersible turbine	Moderate to high	Unlimited	Yes	Yes	
	Mechanical	Moderate to high		Yes	Yes	
	Positive displacement pumps	Moderate to high		Yes	Yes	
Two-pump system	Submersible turbine	Moderate to high	Unlimited	Yes	Yes	Requires significant coproduced water handling capabilities
Skimming units	Floating		Unlimited	Yes	No	
	Suspended		Unlimited	Yes	No	

Others					
Timed bailers		0–100	Yes	No	
Rope skimmers		0–15	Yes	No	Surface water use only
Belt skimmers		0–15	Yes	No	Surface water use only
Vacuum assisted	Low to moderate	0–30	Yes	Yes	
Vapor extraction		0–30	No	No	Product volatility
Biodegradation		0–50	No	Yes	Availability of oxygen and nutrients

[a] Low   = < 10^{-5} cm/s
Moderate = 10^{-5}–10^{-3} cm/s
High   = >10^{-3} cm/s

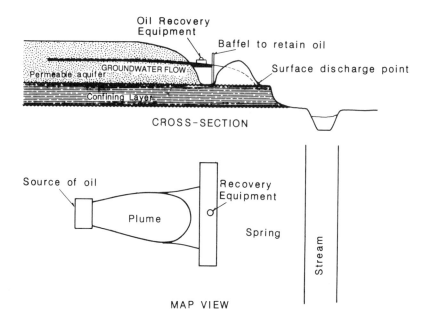

Figure 7.1.   Idealized subsurface hydrogeologic conditions for use of a passive trench system.

7.1.2   Active Systems

The method most often used to improve the performance of trenches is to provide flow enhancement. Typically this involves the use of a pump to lower the fluid level in the trench and thus to increase the hydraulic gradient. An additional benefit is that the continued flow toward the pump also tends to collect the floating product in a smaller area, where it is more easily recovered. Water recovered from the trench may be treated for offsite disposal or reinjected upgradient to further enhance the flow. Increased hydraulic gradient reduces travel time in direct proportion to the slope of the water table. Downgradient migration is also reduced or eliminated because the water flows into the trench from both sides, thus recovering LNAPL product that had migrated immediately downgradient of the trench.

A refinement of trench design is to shorten fluid travel distance by extension of the trench into the plume itself. The use of a dendritic pattern, as shown in Figure 7.2, can assure that flow time is reduced to a practical minimum. Recovery systems of this design are well suited for large open areas with shallow groundwater and flexible air quality regulations.

A rough approximation of the water discharge expected from a pumped trench can be made by applying Darcy's equation:

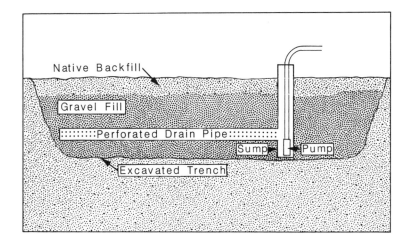

Figure 7.2.   Schematic showing a pumpage-enhanced recovery trench.

$$Q = PIA, \qquad (7.1)$$

where, Q = discharge; I = hydraulic gradient on either side of the trench; P = permeability of soil; and A = area of the trench sides.

For initial calculations, a hydraulic gradient of 0.3 is often used for fine sandy soils. When the hydraulic gradient adjacent to the trench is near this value, the flow velocity in the trench is usually sufficient to bring the floating LNAPL product to the recovery unit.

Variations of linear interceptors include closed trenches that are backfilled with highly permeable material or contain a conduit pipe (classic French drain). When set slightly below the yearly low water table elevation, a trench drain can be an effective interception device. A schematic plan of a trench drain is shown in Figure 7.3.

7.2   WELL-POINT SYSTEMS

Where the LNAPL is located at shallow depth (<20 ft or 6 m) in reasonably permeable material (>EE–5 cm/s), a well-point system may be the appropriate recovery technology. The use of well-points in recovery is similar to construction dewatering. A major advantage to this type of system is that it is possible to lower a shallow fluid surface over a predetermined area and to encourage inflow from adjacent areas at a reasonable cost.

Well-points are commonly small diameter pipes (2–4 in. or 5–10 cm) attached to a short length of well screen (2–3 ft or 60–90 cm), which are installed in sets and are connected to a common suction pump. This type of system is best suited

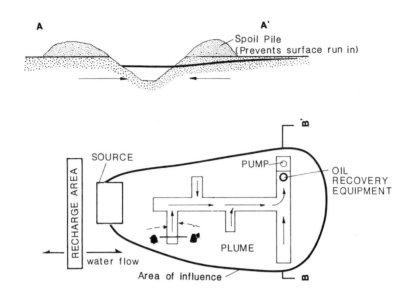

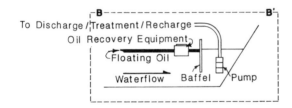

Figure 7.3.   Schematic of a trench drain system.

where the aquifer is under water table conditions. In field practice the maximum height that the fluid can be lifted is a function of the vacuum pump used for recovery. Although some pumps can lift water 20–22 ft (6–7 m), most systems have sufficient leaks to limit the lift to 15–17 ft (4.5 –5 m).

The design of a well-point system for remediation is similar to that of a dewatering system, with a few notable exceptions relating to the nature of the product to be recovered. Prior to the design effort, a hydrogeologic investigation is necessary to assess:

- Geologic and hydrogeologic conditions (grain size distribution, packing density, hydraulic conductivity, etc.)
- Depth to top of fluids
- Product characteristics (density, viscosity, solubility, vapor pressure, etc.)
- Estimated range of fluid level fluctuation during recovery operation

- Aerial extent of floating and dissolved product
- Estimated quantity of recoverable product

Calculation of well-point spacing and the expected pumping rate is usually based on procedures using the Depuit equation, as described in standard texts. The following factors must be considered in addition to standard water pumping considerations:

- The top of the screen must be located at or very near the top of the fluid level during pumping
- If recovery of contaminated water is not a prime concern (after initial drawdown), the screens should be short (<1 m) with closer well spacing to compensate
- Flow meters (or at least site tubes) are helpful to determine if each well is functioning
- Large quantities of volatile product (i.e., gasoline) will reduce the available lift height in direct relation to the vapor pressure

Well-points are typically installed at calculated spacings to cause a limited drawdown over the area necessary to recover the product and contaminated water, but still maintain containment, notably on the downgradient side. A typical well-point is shown in Figure 7.4; a typical well-point installation layout is shown in Figure 7.5.

The use of a well-point system can be highly effective in the remediation of contaminated shallow aquifers, however, several safety hazards are associated with this or any vacuum-lift process. Vacuum increases vaporization, especially when dealing with products similar to gasoline. Discharge of these vapors may not be permitted by regulation. When mixed with air, these vapors can potentially result in explosive mixtures. Air can mix with vapors at the discharge point through leaks in vacuum lines or can be introduced through permeable soil if the fluid level in any well point drops below the top of screen. Because of the potential explosion hazard, it is important to use only certified explosion-proof equipment for pumping or vapor handling.

7.3   VACUUM-ENHANCED SUCTION-LIFT WELL-POINT SYSTEM

The double-diaphragm suction-lift pump LNAPL recovery system is patterned after the concept of a shallow well-point dewatering system commonly used in the construction industry. Well-point dewatering systems typically consist of multiple shallow wells or drive points manifolded to a large capacity suction-lift pump. By pumping at the well-points, overlapping cones of depression are imposed on the water table and a barrier to groundwater is maintained.

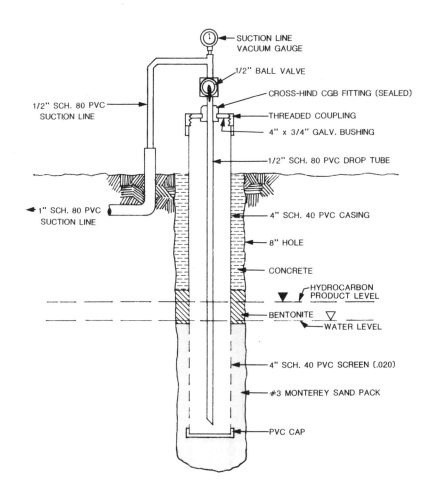

Figure 7.4.   Typical well-point construction.

Adaptation of the well-point dewatering concept to a petroleum hydrocarbon recovery system was initially utilized to inhibit free hydrocarbon seepage to a surface water body. A line of shallow pumping wells was installed adjacent to the surface water shoreline. The overlapping cones of depression at these pumping wells created a groundwater flow boundary and a subsequent LNAPL product flow boundary such that seepage of product to the surface water body was contained. This type of system was then expanded to inland areas and used to create wider zones of pumping influence, thereby producing a general inhibition of product migration and enhancing product recovery from the water table. The well-point system is used to create a flow barrier, as opposed to an open trench collection system, for three primary reasons: (1) open trenches with

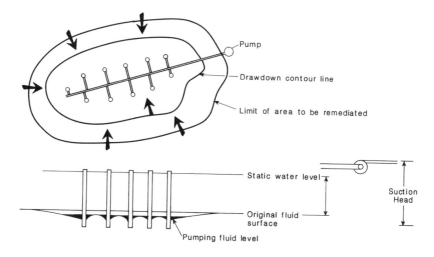

Figure 7.5.   Typical well-point installation layout.

free product can be a fire hazard, (2) site facilities such as underground piping preclude practical trench excavation, (3) well construction minimizes the generation of potentially hazardous soils in contrast to significant volumes potentially generated by trench excavation.

The general recovery system is comprised primarily of the following components:

- Monitoring and recovery wells
- Pumping system
- Product storage and separation system
- Necessary materials such as air and discharge lines

Well spacing is determined based on aquifer pumping tests, which provide data on aquifer permeability and the radius of influence per well at various pumping rates. Wells are placed to provide overlapping cones of influence within the constraints of site specific factors such as interference with buildings, aboveground tankage and retaining walls, subsurface piping, etc.

Modifications to the well-point system for use in LNAPL recovery includes use of explosion-proof pumping equipment and specific recovery well alterations. The bore holes for recovery well installation can be drilled by either machine-driven, hollow-stem auger or by hand auger in less accessible areas. Two basic alterations from standard monitoring well design are made during the construction of the recovery well: (1) screen length and placement and (2) sealed well heads. To maximize the efficiency of the recovery wells, the screened interval is generally placed beginning at the bottom of each recovery well. However, the top of the screen is placed up to 5 ft below the water table (or free hydrocarbon) surface under nonpumping conditions. Under pumping conditions, the water level (product level) is drawn down to the point where LNAPL can enter the well. The deeper screened interval allows for minimization of vacuum loss, since less screened area is in contact with the unsaturated zone as the water level is lowered by pumping.

The well head on each recovery well is sealed to promote additional fluid entry into the well. Sealing the well head allows for a vacuum to be created within the well casing using suction-lift pumping equipment. The generation of a vacuum increases the effective difference in pressure head between the inside of the casing and the adjacent formation, thereby stimulating additional drawdown within the pumped well. As the amount of drawdown (or simulated drawdown) increases, the flow of product and groundwater to the well increases. With a well that is relatively shallow in depth, the additional simulated drawdown induced by vacuum enhancement results in increased flow to the well in the absence of additional available drawdown.

Components of the pumping system reflect both practical aspects related to characteristics of the site facilities and the subsurface hydrogeologic conditions encountered. The pumps selected are pneumatic, double-diaphragm, suction-lift pumps. The advantages of these pumps include:

- Suction-lift pumps are pneumatically operated and thus are intrinsically safe
- Air supply typically is readily available from existing on-site, large-capacity compressors
- Pumps are self-priming to a depth of 22 ft and can be effective to depths of 27 ft providing suction is not broken
- Pumps provide pressurized pumping from the well to the storage/separation tank in addition to suction of fluids from the wells
- In the event a well(s) is pumped dry, the pumps can pump air without damage. Pumping air also induces the vacuum-enhanced fluid entry into the wells

The pumps utilized are commercially available and constructed of cast iron or stainless steel with Buna-N rubber diaphragms. Each pump is conventionally manifolded to withdraw groundwater from up to four recovery wells. Discharge is pumped into a piping system leading to a storage tank for product/water separation and product storage. Pump controls include intake and discharge valves, a compressed air dryer, a compressed air supply regulator, and a pump/oiler. A sample discharge port was installed for monitoring the pumping rate and assessment of product volume discharged by each pump. A schematic of pump controls is presented in Figure 7.6.

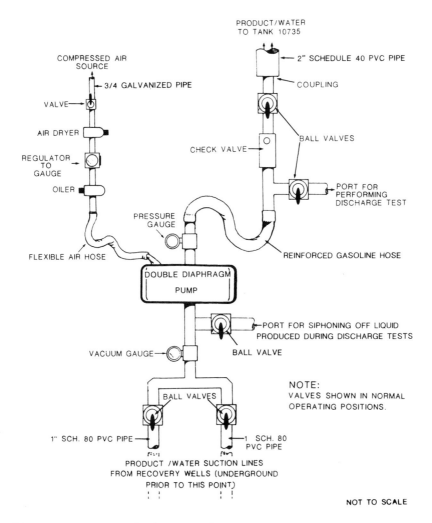

Figure 7.6.   Schematic of double-diaphragm pump controls.

The number of wells manifolded to an individual pump primarily reflects the depth to groundwater, but is also in part dependent on the pump size. In low permeability formations, wells are typically pumped dry. As long as the pumps sustain sufficient vacuum to all wells at the intake depth, product eventually enters the well and is evacuated. In higher permeability formations, fluid is pumped consistently from each well at a maximum rate. In both low and high permeability formations, individual pumps work harder to pump from greater depths. Decreasing the number of wells per pump maximizes the pumping response of individual wells. Although not specifically quantified, this is inferred to be the result of a single pump not being capable of applying the necessary vacuum to the piping and well volume of multiple wells under deeper groundwater conditions. Where lift requirements are at or near the suction-lift capability of the pump (about 22 ft), only one well per pump is used. For shallow groundwater conditions of 10 ft or less, up to four wells have been manifolded to one pump with favorable results.

The fluid recovered from the wells is a mixture of groundwater and petroleum hydrocarbon product. Since water and LNAPL are immiscible, the two liquids will separate in an open container, such as a storage tank. The free hydrocarbon, having a lesser specific gravity than water, floats to the top, while water descends to the bottom of the tank. Depending on the configuration of the storage tank, the water is pumped or allowed to drain out the bottom of the tank. Most petroleum-handling facilities maintain existing tanks for routine product separation and storage.

Water discharge is typically handled by the existing treatment and discharge facilities, or treated by carbon filtration prior to discharge where the water volume has been minimal. LNAPL product is periodically pumped out of the tank and transported by truck to a recycling facility. Periodic tank gauging and product transfers avoid the accumulation of product to the point where it could be inadvertently discharged with the water.

For effective LNAPL recovery, the suction-lift pumping system requires periodic monitoring, and operation and maintenance of the equipment. Since the pumps have a minimum of moving parts and control is accomplished by valving of the air supply, the maintenance of the equipment is minimal.

To assess the effectiveness of the system and to maximize recovery, data collection is periodically performed. Data routinely collected includes monitoring of pumping rates at each pump and calculating the percent product of the discharge in total volume to estimate the volume of product recovered. Storage tank gauging is performed to monitor the overall product volumes recovered. Periodic gauging of monitoring wells installed throughout the recovery area is also conducted to monitor reductions in the apparent product thicknesses over time.

Gauging is performed using an electric petroleum hydrocarbon product and water level resistivity probe or steel tape with water-and oil-finding paste. For each gauging event, the depth to product and depth to water is measured for each monitoring well. These measurements are entered into a computer database for subsequent data compilation and analysis. The corrected groundwater elevation from depth to fluid measurements is calculated based on well head elevations and the specific gravity of the product as determined through laboratory analysis.

7.4   ONE-PUMP SYSTEM

At sites where the water table is below the suction-lift depth, the use of some variety of down-hole pumping unit is necessary. When a single pump system is

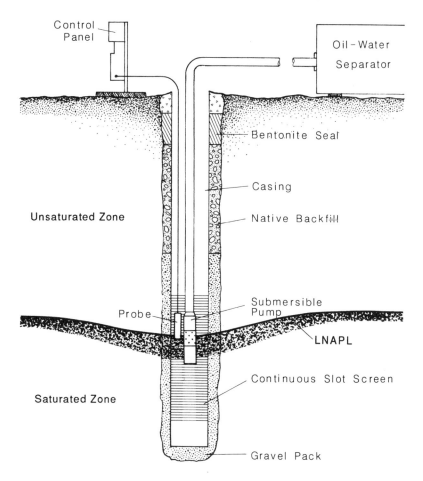

Figure 7.7.   Schematic of a single-pump system using one recovery well.

used, water and/or product is delivered to the surface for separation or treatment, as shown in Figure 7.7. Several types of single pump systems are in common usage. The only feature shared by all of these is a single intake port. General types include:

- Submersible turbine (electric)
- Mechanical-lift pumps

7.4.1   Submersible Turbine

Submersible turbine pumps are a variety of vertical turbine pumps that have the motor attached below the pumping unit. Water passes in through an intake port located between the motor and bowl assembly, then upward through the bowl stages up to the surface through the pump pipe. Electric power is supplied to the motor by specially insulated wires. This type of pump is manufactured for water supply and oil well usage by a wide variety of manufacturers in sizes ranging from one-third to several hundred horsepower, and is constructed of a variety of materials suited for many chemicals.

After a recovery-remediation well has been constructed and tested for well capacity, an appropriate submersible pump may be installed. Materials of construction must be compatible with the fluids to be pumped (water, oil, silt/ sand). Most pumps intended for domestic usage contain internal parts that are not resistant to the chemicals that may be encountered in remediation (either dissolved or LNAPL). Motor seals, impellers, check valves, and electric lead wires are the most common components to fail. When the pumped fluid is primarily a petroleum product, the pitch angle of the impellers must be compatible with the viscosity of the fluid. Also, the pump manufacturer should warrant that the pump is explosion proof if there is any possibility of pumping vapors or LNAPL hydrocarbon product.

Installation of submersible pumps in recovery wells requires specific attention to the intake setting. If the intent of the recovery is to reclaim floating product, the intake must be set at the elevation of the interface during pumping. Prior to installation of the pump, the well must be thoroughly tested to determine the specific capacity so that the pumping level for each pumping rate can be calculated. The pump should be supported by cable, rigid pipe, or other support that is attached to a hoisting mechanism, which will allow the pump to be raised or lowered to accommodate fluctuations of the interface surface. Regular adjustments are necessary to maintain optimum performance, i.e., maximum product recovery with minimum water retrieval.

Fluid level monitoring is a necessary precaution to protect the pump from pumping dry. Level control can be accomplished by the use of a simple float switch, an air-bubbler pressure switch, amperage control, an interface probe, or other device. Usually, these controls are attached to an electronic timing unit that is adjusted to stop the pump for a specific period of time, which allows the well to recover, and then restarts the pump. Although this type of control can be used to operate the recovery system on a cyclic (on-off) manner, it is not as efficient as continual pumping and is very hard on the pump starting mechanism.

A recent innovation to improve recovery functions is a variable-speed submersible pump. When used with electronic controls, it is possible to select a fluid surface elevation and to adjust the pumping rate periodically to maintain that fluid level at that elevation. Another adaptation of existing control technology uses an electrically adjustable valve to throttle the pump output. As the fluid level declines, the throttle restricts flow to prevent a further decline.

7.4.2   Positive Displacement Pumps

The use of traditional rod and piston pumps continues at many LNAPL recovery locations, particularly at refineries and distribution terminals. These units are usually powered by single-speed electric motors and have adjustable stroke lengths to control the pumping rate. When installed with the intake set at "optimum" pumping depth, they function fairly well. The primary advantage of rod and piston pumps is that the smooth slow stroke rate can pump mixtures of product and water without creating a significant emulsion.

Several disadvantages should be carefully considered prior to their installation. Standard construction of these units consists of a fixed length column pipe that contains the sucker rods. The intake is attached at the bottom, and is at a set distance from the pump jack. Raising or lowering of the intake to accommodate fluctuations of the fluid surface requires either adding/subtracting lengths of the pump pipe and sucker rod or changing the height of the pump jack. Both of these options require significant field effort. Another mechanical difficulty is that the reciprocating stroke action is strenuous on the moving parts. The structural components necessary for these units to provide satisfactory service causes them to be uneconomical compared to other recovery options. Finally, the packing seals around the rods at the surface have a great tendency to leak, causing product to be discharged onto the surface casing. At remediation sites, this does not usually present an appropriate visual image.

7.5   TWO-PUMP SYSTEM

The recovery system most often used at sites with significant quantities of recoverable LNAPL product is the two-pump system. A submersible water pump installed below the lowest possible probable interface level is used to create a drawdown cone of depression, while a second pump is suspended with its intake port located at the oil/water interface. A typical two-pump system installation is shown in Figure 7.8.

The lower pump is controlled to produce sufficient water to cause a cone of depression extending outward to intercept and retrieve the LNAPL product. While the water pump operates continually, the upper product pump cycles on and off to recover the product as it accumulates.

Automatic interface detection probes attached near the intakes of both pumps provide an operating "logic". The upper probe is adjusted to detect both air/product and product/water interfaces to assure that the pump only recovers product and does not run dry. The lower probe is set to detect the presence of a product/water interface and stop that pump before it discharges any product. If for some reason the upper pump should not work and product accumulates,

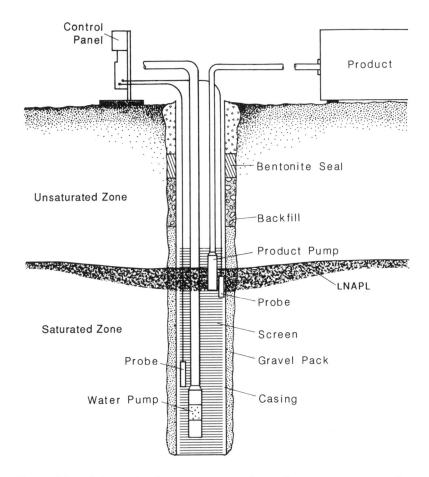

Figure 7.8.   Schematic of a two-pump system using one recovery well.

forcing the interface down toward the lower pump intake, the entire system ceases to operate.

There are numerous advantages to a two-pump system. Product is separated in the well, reducing the need for aboveground separators. In several situations, the product can be recycled without further treatment. Because the mixing of product and water in the well are minimized, soluble components are not added to the pumped water. The most important advantage is that the system is fully automatic and can be operated continuously with minimal adjustment after start-up.

While this system is the preferred choice for many situations, it is commonly overused, and there are several disadvantages that must be considered. Because all of the operating components are down the well, when they require occasional cleaning or adjustment they must be brought to the surface. Also, a larger diameter well is required to contain all of these components. It is common to specify an 8–12 in. (minimum) well to assure that there is sufficient space for operation and access. Routine maintenance is essential, as interface probes tend to become coated with oil which interferes with their function.

The initial start-up adjustment of two-pump systems requires extreme care. The water pumping rate and pump setting must be adjusted to set the product/water interface at nearly a constant level while maintaining the necessary drawdown to assure the preservation of the capture zone. Regular gauging must be made during the first few days of operation to assure the proper settings. Particular attention must be used in nonhomogenous or low-yield aquifers because stabilization of pumping levels may be difficult. Experienced staff are very important for start-up and maintenance of this type of system.

7.6   SKIMMING UNITS

The recovery of relatively thin, but persistent thickness of LNAPL from a well or open trench system can be accomplished with a skimming unit. Most "skimmers" are designed to recover product thickness of a few millimeters (fractions of an inch) at fairly low rates (.2–3 gpm). These units are used primarily in wells, however, they are easily adapted for use in open fluid surfaces (trench or pond). They may be used in conjunction with water-level depression pumps.

7.6.1   Floating Skimmers

Floating skimmers float freely on the fluid surface to skim floating product and pump it to the surface. Most manufacturers provide an oleophilic filter to limit the quantity of water recovered. Floating skimmers are manufactured in sizes ranging from 2 to 28 in. in diameter. The smaller units are usually attached to a surface-suction pump, while the larger units usually have a self-contained pump.

The primary advantage of floating skimmers is that they are constantly in contact with the water surface. If the fluid level changes, they adjust automatically. Also, these units are fairly easily installed and can be moved from location to location with minimal effort.

The major disadvantage is that they tend to become stuck in wells that are too small, or that contain other piping and gauging tubes, or do not have a smooth surface on the inside of the well screen. In wells, fixtures must be properly installed to allow adequate free motion without restriction. It is usually helpful to install additional piping, gauges, wires, or other equipment enclosed in a separate conduit attached to the interior casing wall.

7.6.2   Suspended Skimmers

Another variety of low rate product recovery pump is a pneumatic operated chamber pump, which is suspended with its intake set at the product/water interface to recover mostly product. Most of these pumps have diameters of 1.5–4 in. (although larger units are available) with fixed chamber volumes that may be either top filling or bottom filling (for DNAPL), and are controlled by a timing control. By varying the pulse cycle, pumping rates can be set at .01–4 gpm for larger units. Many commercial brands can be adapted to this usage (or for groundwater monitoring wells).

The bladder pump is similar in concept with a flexible membrane bladder attached to the air line inside the chamber. A bladder prevents operating air from contacting pumped fluids; therefore, when the air is discharged to the atmosphere it does not contain vapors. A schematic of an open-chambered skimmer pump is shown in Figure 7.9.

Suspended pneumatic skimmers recover product where sufficient product is accumulated to allow timed recovery. They are installed as single-pump recovery units and also as the product pump for two-pump recovery units. After the cycle time is adjusted for a semi-steady-state pumping situation, they are easily controlled. Raising and lowering the intake to accommodate fluctuating fluid levels or adjusting the time cycle as product accumulates are the primary adjustments necessary.

7.7   OTHER RECOVERY SYSTEMS

Several ingenious manufacturers have developed recovery units that are particularly effective in specific situations. The alternative recovery technologies presented in the following paragraphs are presented to acquaint the reader with the general type of equipment commercially available and to stimulate the development of additional designs.

7.7.1   Timed Bailers

The ancient technique of using a bailer bucket to recover fluid from a well has been adapted to recover LNAPL product. Commercial units available are constructed to lower a sealed-bottom bailer into a well until the top of the bailer is slightly below the fluid level. When the bailer fills, it is retrieved to the surface, where it is discharged into an oil/water separator. After a set period of time the cycle is repeated. When properly adjusted, and with continued attention, these units can be effective in the removal of accumulated product from wells. Manufactured units are usually available for wells of 2 in. diameter and greater.

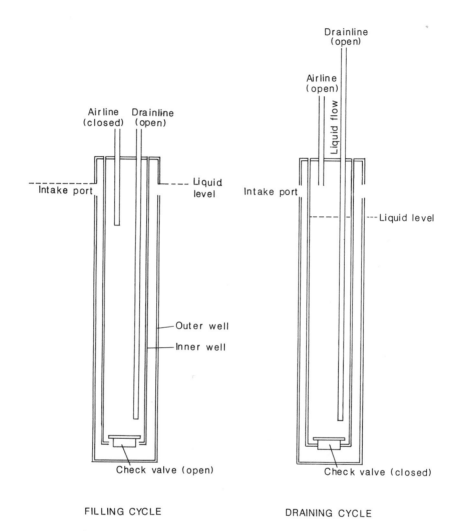

Figure 7.9. **Schematic of an open-chambered skimmer pump.**

The advantage of the automatic bailer bucket is that commercial units are self-contained and readily mobile. They can be set up in a short period of time and can be adjusted to be highly effective.

The primary disadvantage is that if these bailers are adjusted to remove only the product layer, they do not create a drawdown cone. Therefore, unless the product layer is relatively thick and quite fluid, the recovery rate tends to be slow. Also, use of bailers in deeper wells requires significant bailer travel time, which can diminish the effectiveness.

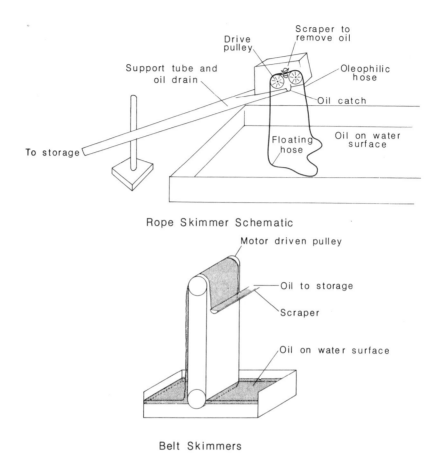

Figure 7.10. Schematic of a rope (a) and belt (b) skimmer system.

7.7.2   Rope and Belt Skimmers

Rope and belt skimmer units are well adapted to removing thin layers of floating LNAPL product from open trenches and other large surface areas (ponds or even manholes). Both systems rely upon continuous rotation of a flexible closed loop tube (or belt) constructed of an oleophilic material (Figure 7.10). The automatic unit draws the oil-covered tube (or belt) through scrapers and returns it to the surface to gather more oil.

Where a tube is used, an excess length is allowed to float on the water surface snaking around the recovery area. The skimmer is usually attached to a cantilever boom that extends out over the fluid surface. The hollow boom also serves as the conduit that conducts the product to storage containers.

Belt units are constructed such that the continuous belt is suspended between a driven pulley above the fluid to a lower portion immersed below the fluid surface. As the belt descends into the liquid, the floating oil adheres to both sides of the belt. Wipers attached to the upper unit remove the oil, which drains into a trough that conveys it to storage containers.

The advantage of either rope or belt skimmers is that they do not require significant operating attention. They can be left unattended (except for routine maintenance) for extended periods of time. Also, either unit can recover thin layers of floating product without sophisticated instrumentation.

The primary disadvantage is that the recovery capacity is limited to a few hundred gallons per day per unit. More viscous oil products attach more effectively to the oleophilic materials. Less viscous products are recovered less effectively.

7.7.3   Vacuum-Assisted Systems

The recovery of LNAPL product from a low permeability aquifer or a more viscous product from a moderately permeable formation is often difficult due to the limited hydraulic gradient that is developed by conventional well pumping systems, especially in limited thickness formations. The application of a vacuum to the casing can often increase more viscous product formation by a significant amount.

Several designs have proven to be effective. When the fluid is within suction range, a well equipped with a "dip tube" and casing seal can be used. When fluid occurs at greater depths, the low pressure may be created by a separate vacuum pump or by an in-well eductor system.

Both systems described above provide potentially increased product production, but at the expense of producing both product and water. Also, significant quantities of vapors are usually produced.

7.7.4   Vapor Extraction and Biodegradation

Volatile LNAPL product in the subsurface vaporizes from the water table into air-filled pore spaces until the vapor-liquid equilibrium concentrations are established. If the soil is naturally permeable above the fluid surface to the soil surface, a concentration gradient is established. Eventually, most of the volatile mass can be transferred from the floating LNAPL liquid to the atmosphere, minus that portion retained by sorption in the soil.

Several attempts have been reported by anonymous recovery teams to recover LNAPL product by the use of vacuum wells set above the fluid surface. These efforts have attained marginal success; however, several factors have been seen to interfere. Vacuum wells set in the vadose zone tend to encourage air flow from the surface downward through the soil, as well as to extract petroleum vapors. The result is often the enhancement of biological degradation near the wells, which tends to cause the well screens to become plugged with biomass. Routine maintenance of the wells is required to keep them functioning properly.

The major disadvantage to recovery of LNAPL product by vapor extraction is the large quantity of energy required to maintain rapid volatilization of the product and air flow to transport the diluted vapors to the surface. The remediation time for this type of system tends to be quite long, often in terms of years.

A few causes of LNAPL product removal by natural biodegradation have been documented (in specific physical settings). In most of these situations, the water table fluctuations are relatively large and are within a few meters of the surface. Naturally occurring bacteria attached to the permeable soil structure act as a fixed-film reactor. When the water table rises, it brings the product to the biomass. As it falls again, some of the free product remains sorbed to the biomass-soil structure in contact with air that is induced into the newly emptied pore spaces. After several cycles of treatment, a significant quantity of LNAPL product can be degraded.

Reliance upon natural processes to remediate aquifers is occasionally an effective technique. However, if an aquifer is to be restored in a reasonable period of time, all of the necessary conditions must be met. Essential nutrients, sufficient oxygen, lack of biotoxins, and hydraulic containment are required to accomplish the destruction (or reformation) of the product. Few aquifers have the proper balance of these conditions to complete the process before the contamination spreads to areas beyond the site.

REFERENCES

1.   American Petroleum Institute (API), 1980, The Migration of Petroleum Products in Soil and Groundwater - Principles and Countermeasures. API Publication No. 1628.

2.   American Petroleum Institute (API), 1988, Phase Separated Hydrocarbon Contaminant Modeling for Corrective Action: American Petroleum Institute, Health Environmental Sciences Department, Publication No. 4474, 60 p.

3.   Bellandi, R., 1988, *Hazardous Waste Site Remediation, The Engineer's Perspective*: Van Nostrand Reinhold, New York, New York, 422 p.

4.   Blake, S. B., and Gates, M. N., 1986, Vacuum-Enhanced Hydrocarbon Recovery: A Case Study. In Proceedings of the National Water Well Association Conference on Petroleum Hydrocarbons and Organic Chemicals in Groundwater: Prevention, Detection and Restoration, November, 1986, p. 709-721.

5. Blake, S. B. and Gates, M. M., 1986, Vacuum-Enhanced Hydrocarbon Recovery: A Case Study. In Proceedings of the Second Annual Hazardous Materials Conference West, Tower Conference Management Company, Wheaton, IL, p. 17-25.

6. Burke, M. R. and Buzea, D. C., 1955, Unique Clean-up of Hydrocarbon Product from Low Permeability Formation Progressing in St. Paul, Minnesota: *Hydrological Science and Technology*, Vol. 1, No. 1, p. 53-58.

7. Conner, J. A., Newell, C. J., and Wilson, D. K., 1989, Assessment, Field Testing, and Conceptual Design for Managing Dense Non-Aqueous Phase Liquids (DNAPL) at a Superfund Site: In Proceedings of the National Water Well Association and American Petroleum Institute Conference on Petroleum Hydrocarbons and Organic Chemicals in Groundwater: Prevention, Detection and Restoration, November, 1989, p. 519-533.

8. Farmer, V. E., 1983, Behavior of Petroleum Contaminants in an Underground Environment. In Proceedings of Groundwater and Petroleum Hydrocarbons - Protection, Detection, Restoration. PACE, Toronto, Ontario.

9. Lindsly, B. E. and Berwald, W. B., 1930, Effect of Vacuum on Oil Wells: *U. S. Bur. Mines Bull.*, No. 322, 133 p.

10. McKee, J. E., Laverty, F. B., and Hertel, R. M., 1972, Gasoline in Groundwater. *Journal of Water Pollution Control Federation*, Vol. 44, p. 293-302.

11. Powers, J. P., 1981, *Construction Dewatering*. John Wiley and Sons, New York, N.Y., 484 p.

8 AQUIFER RESTORATION

"Effectiveness of most recovery operations is limited not by available technology, but rather the inability to handle coproduced water"

8.1 INTRODUCTION

During the process of recovering LNAPL that has accumulated in the subsurface, groundwater contaminated with dissolved fractions is most often coproduced. Depending on the size of the facility and the scale of the recovery project, the amount of groundwater coproduced can possibly exceed 1000 gal/min. Handling of these volumes of coproduced water can be very expensive if treatment of the water is required prior to disposal or reinjection.

The type of treatment and disposal method required for coproduced water will depend primarily on the type of specific contaminants, the intended water use, and the treatment levels required. Treatment systems may be relatively simple in the rare case of a single chemical contaminant, or extremely complex for cases involving numerous contaminants. Disposal options range from direct discharge and reinjection without treatment, to discharge and reinjection after treatment, or to reinjection to recharge basins and allowing to percolation.

Normally, treatment of coproduced groundwater during hydrocarbon recovery operations will include, as a minimum, oil/water separation and the removal of dissolved hydrocarbon fractions (i.e., benzene, toluene, and total xylenes). In addition, removal of inorganic compounds and heavy metals (i.e., iron) is often required. Iron contamination, which is a common dissolved constituent in groundwater, for example, may require treatment prior to downstream treatment processes to prevent fouling problems in air stripping systems. Heavy metals removal is normally accomplished by chemical precipitation.

The technologies that have been found to be applicable for dissolved hydro-

153

carbons include air stripping, activated carbon adsorption, biological treatment, and various combinations of these technologies. In some cases, especially under low groundwater pumping flow rate requirements, existing refinery wastewater treatment facilities may be used for groundwater treatment.

An overview of treatment technologies, along with site-specific considerations, and cost comparisons for various approaches that have been utilized for coproduced groundwater treatment are discussed below. In addition, disposal options are presented with an emphasis on the benefits of reinjecting untreated groundwater during hydrocarbon recovery.

8.2   OIL/WATER SEPARATION TECHNOLOGIES

Removal of LNAPL can present problems, particularly if emulsion is involved. Emulsion is an intimate mixture of two liquids not miscible with each other, as oil and water. Water-in-oil emulsions have water as an internal phase and oil as the external phase, whereas oil-in-water emulsions reverses the order. Oil/water separation is required prior to downstream treatment processes. An overview of specific oil removal technologies is presented below.

8.2.1   Gravity Separation

Gravity oil separation equipment, which includes API separators, tanks with skimming, and various skimming clarifier designs, is efficient in removing large amounts of free oil from coproduced groundwater. While gravity separation is effective in removing free and unstable oil emulsions, soluble oil fractions and many emulsions are not removed by this type of separation equipment. Primary gravity separation is the most economical and efficient way to remove large quantities of LNAPL hydrocarbon. The effluent oil concentrations from gravity separators in refineries generally range from 30 to 150 mg/l, although deviations on either side of this range may occur. The removal efficiency for all gravity separation equipment is a function of temperature and density differences between the oil and water.

8.2.2   Dissolved Air Flotation

Dissolved air flotation (DAF) is another process used in removing "oil and grease" constituents. In this process, groundwater, or some fraction thereof, is saturated under pressure with a gas (usually air). Upon release of this pressure, the air in excess of the atmospheric saturation concentration is released from solution, forming bubbles of approximately 30–120 μm in diameter. The bubbles form on the surfaces of the suspended or oily materials or are attached to the particles by interfacial attraction. Consequently, an aggregate is formed with an average density substantially less than that of water causing the aggregate to rise. Dissolved air flotation can be used by itself but, for maximum effectiveness, is used with chemical coagulation and flocculation. The chemical treatment aspects of DAF operation are extremely important, particularly when colloidal or emulsified oil components are present. Coagulants, such as lime, alum, ferric salts, or polyelectrolytes, are used to improve floc formation and to provide good separation. They can be injected at several points depending on the flotation process and the chemical used. Effluent oil concentrations averaging less than 20 mg/l are typical from well-operated DAF units.

8.2.3   Chemical Coagulation-Flocculation and Sedimentation

This process includes the addition of chemical coagulants to produce microflocs by coagulation, followed by flocculation to produce larger particles that can be removed by sedimentation. Upflow clarifiers have been used in lieu of the conventional flocculation-sedimentation system with chemical addition. In an upflow clarifier, the floc formed is removed in the sludge blanket as the water flows upward. Chemicals such as lime, alum, and polyelectrolytes are added before the groundwater enters the clarifier.

8.2.4   Coalescence

Fibrous bed coalescers generally have a fixed filter element constructed of fiberglass or other materials that acts to coalesce the oil droplets and to break emulsions. The coalesced oil droplets released from the filter are readily separated downstream by gravity. Coalescence in a fibrous bed coalescer involves three steps:

1. Interception of fine droplets by fibers
2. Attachment of droplets to the fibers or to retained droplets
3. Release of enlarged droplets from the fibers

Demulsification by induced coalescence requires the rupture of the protecting film as the emulsion flows through the small passages in the fibrous media. Coalescence of the dispersed phase is then possible because of the preferential oil wetting characteristics of the media surface.

8.2.5   Membrane Processes

Membrane processes such as ultrafiltration or reverse osmosis have been proposed as oil removal processes. Laboratory tests have indicated favorable oil removal, although relatively low flux rates, membrane fouling, and membrane-life problems have presented concerns for the practical application of membrane processes to oil removal.

8.2.6   Biological Processes

Biological treatment processes have limitations in their applicability to LNAPL removal. Microorganisms are efficient in oxidizing most soluble organic compounds, including some dispersed or emulsified oils. Large amounts of LNAPL must be avoided, as they coat the biological floc, interfere with efficient oxygen transfer within the biomass, and produce oily sludge scums.

8.2.7   Carbon Adsorption

Activated carbon adsorption has very limited use in the removal of LNAPL. Adsorption is primarily effective for removal of low levels of soluble hydrocarbons. Groundwater applied to activated carbon adsorption units must be pretreated to prevent clogging and coating of the activated carbon with free oil. If the activated carbon adsorption units are not adequately protected, the units will have to be backwashed frequently and the activated carbon will have to be replaced at an acceptable frequency.

8.3   INORGANICS REMOVAL

Chemical addition for the removal of inorganic compounds is a well established technology. There are three common types of chemical addition systems that depend upon the low solubility of inorganics at a specific pH. These include the carbonate system, the hydroxide system, and the sulfide system. In reviewing the basic solubility products for these systems, the sulfide system removes the most inorganics, with the exception of arsenic, because of the low solubility of sulfide compounds. This increased removal capability is offset by the difficulty in handling the chemicals and the fact that sulfide sludges are susceptible to oxidation to sulfate when exposed to air, resulting in resolubilization of the metals. The carbonate system is a method that relies on the use of soda ash and pH adjustment between 8.2 and 8.5. The carbonate system, although workable in theory, is difficult to control. The hydroxide system is the most widely used inorganics/metals removal system. The system responds directly to pH adjustment, and usually uses either lime (CaOH) or sodium hydroxide (NaOH) as the chemical to adjust the pH upwards. Sodium hydroxide has the advantage of ease in chemical handling and in producing low volume of sludge. However, the hydroxide sludge is often gelatinous and difficult to dewater.

Chemical precipitation can be accomplished by either batch or continuous flow operations. If the flow is less than 30,000 gpd, a batch treatment system may be the most economical. In the batch system, two tanks are provided, each with a capacity of one day's flow. One tank undergoes treatment while the other tank is being filled. When the daily flow exceeds 30,000 gpd, batch treatment is usually not feasible because of the large tankage required. Continuous treatment may require a tank for acidification and reduction, then a mixing tank for chemical addition, and a settling tank.

The important design factors that must be determined for a particular water during treatability studies include:

- Best chemical addition system
- Optimum chemical dose
- Optimum pH conditions
- Rapid mix requirements
- Flocculation requirements
- Sludge production
- Sludge flocculation, settling, and dewatering characteristics

Laboratory-scale test procedures consisting of jar test studies have been used for years, and the test methodology developed is such that full-scale designs can be developed from these studies with a high degree of confidence.

Chemical precipitation has traditionally been a popular technique for the removal of heavy metals and other inorganics from wastewater streams. However, a wide variety of other techniques also exist. For example, ion exchange, reverse osmosis, evaporation, freeze crystallization, electrodialysis, cementation, catalysis, distillation, and activated carbon have all been used for inorganics removal.

8.4   ORGANICS REMOVAL

The primary concern with coproduced water during LNAPL recovery operations will normally be removal of the dissolved fractions of hydrocarbons. As previously indicated, there are many applicable treatment technologies available for the removal of dissolved hydrocarbons. The commonly used processes are discussed in the following sections.

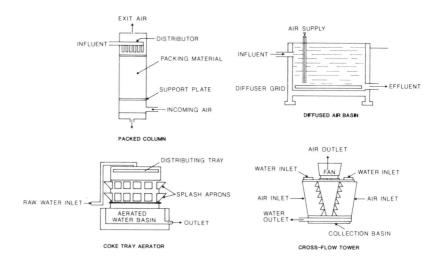

Figure 8.1.   Schematic showing conventional air stripping equipment configuration (after Knox et al., 1986).

8.4.1   Air Stripping Processes

Air stripping and various combinations, such as steam and high-temperature air stripping, have been successfully used for removing volatile organic hydrocarbons from coproduced groundwaters. There are four basic equipment configurations used for air stripping. These include diffused aeration, counter-current packed columns, cross-flow towers, and coke tray aerators, as shown in Figure 8.1. Diffused aeration stripping uses aeration basins similar to standard wastewater treatment aeration basins. Water flows through the basin from top to bottom with the air dispersed through diffusers at the bottom of the basin. The air-to-water ratio is significantly lower than in either the packed column or the cross-flow tower.

In the counter-current packed column, water containing one or more impurities is allowed to flow down through a column containing packing material with air flow counter-current up through the column. In this way the contaminated water comes into intimate contact with clean air. Packing materials are used that provide high void volumes and high surface areas. In the cross-flow tower, water flows down through the packing as in the counter-current packed column; however, the air is pulled across the water flow path by a fan. The coke tray aerator is a simple, low-maintenance process. The water being treated is allowed to trickle through several layers of trays. This produces a large surface area for gas transfer.

The counter-current packed tower appears to be the most appropriate equipment configuration for treating contaminated groundwaters for the following reasons:

- It provides the most liquid interfacial area
- High air-to-water volume ratios are possible due to the low air pressure drop through the tower
- Emission of stripped organics to the atmosphere may be environmentally unacceptable; however, a counter-current tower is relatively small and can be readily connected to vapor recovery equipment

The design of an air stripping process for stripping volatile organics from contaminated groundwater is accomplished in two steps. The cross-sectional area of the column is determined and then the height of the packing is determined. The cross-sectional area of the column is determined by using the physical properties of the air flowing through the column, the characteristics of the packing, and the air-to-water flow ratio. A key factor is the establishment of an acceptable air velocity. A general rule of thumb used for establishing the air velocity is that an acceptable air velocity is 60% of the air velocity at flooding. Flooding is the condition in which the air velocity is so high that it holds up the water in the column to the point where the water becomes the continuous phase rather than the air. If the air-to-water ratio is held constant, the air velocity determines the flooding condition. For a selected air-to-water ratio, the cross-sectional area is determined by dividing the air flow rate by the air velocity. The selection of the design air-to-water ratio must be based upon experience or pilot-scale treatability studies. Treatability studies are particularly important for developing design information for contaminated groundwater.

8.4.2   Carbon Adsorption

Adsorption occurs when an organic molecule is brought to the activated carbon surface and held there by physical and/or chemical forces. The quantity of a compound or group of compounds that can be adsorbed by activated carbon is determined by a balance between the forces that keep the compound in solution and the forces that attract the compound to the carbon surface. Factors that affect this balance include:

1. Adsorptivity increases with decreasing solubility.
2. The pH of the water can affect the adsorptive capacity. Organic acids adsorb better under acidic conditions, whereas amino compounds favor alkaline conditions.

3. Aromatic and halogenated compounds adsorb better than aliphatic compounds.
4. Adsorption capacity decreases with increasing temperature, although the rate of adsorption may increase.
5. The character of the adsorbent surface has a major effect on the adsorption capacity and rate. The raw materials and the process used to activate the carbon determine its capacity.

When activated carbon particles are placed in water containing organic chemicals and mixed to give adequate contact, the adsorption of the organic chemicals occurs. The organic chemical concentration will decrease from an initial concentration of C_o, to an equilibrium concentration of C_e. By conducting a series of adsorption tests, it is usually possible to obtain a relationship between the equilibrium concentration and the amount of organics adsorbed per unit mass of activated carbon. The Freundlich isotherm and the Langmuir isotherm are most often used to represent the adsorption equilibrium.

From an isotherm test it can be determined whether a particular organic material can be removed effectively. It will also show the approximate capacity of the carbon for the application and provide a rough estimate of the carbon dosage required. Isotherm tests also afford a convenient means of studying the effects of pH and temperature on adsorption. Isotherms put a large amount of data into concise form for ready evaluation and interpretation. Isotherms obtained under identical conditions using the same contaminated groundwater for two or more carbons can be quickly and conveniently compared to determine the relative merits of the carbons.

Activated carbon adsorption may be accomplished by batch, column, or fluidized-bed operations. The usual contacting systems are fixed bed or counter-current moving beds, as shown in Figure 8.2. The fixed beds may employ downflow or upflow of water. The counter-current moving beds employ upflow of the water and downflow of the carbon, since the carbon can be moved by the force of gravity. Both fixed beds and moving beds may use gravity or pressure flow.

In a typical fixed-bed carbon column, the column is similar to a pressure filter and has an inlet distributor, an underdrain system, and a surface wash. During the adsorption cycle the influent flow enters through the inlet distributor at the top of the column, and the groundwater flows downward through the bed and leaves through the underdrain system. The unit hydraulic flow rate is usually 2–5 gpm/ft^2. When the head loss becomes excessive due to the accumulated suspended solids, the column is taken off line and backwashed.

In a typical counter-current moving-bed carbon column employing upflow of the water, two or more columns are usually provided and are operated in series. The influent contaminated groundwater enters the bottom of the first column by means of a manifold system that uniformly distributes the flow across the

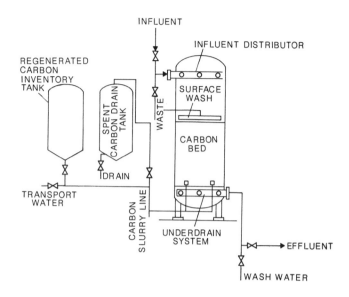

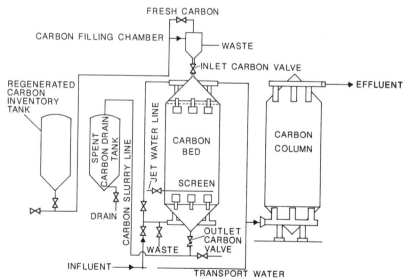

Figure 8.2. Schematic showing fixed-bed and moving-bed adsorption system (after Knox et al., 1986).

bottom. The groundwater flows upward through the column. The unit hydraulic flow rate is usually 2–10 gpm/ft^2. The effluent is collected by a screen and manifold system at the top of the column and flows to the bottom manifold of the second column. The carbon flow is not continuous but instead is pulsed.

The fluidized bed consists of a bed of activated carbon. The water flows upward through the bed in the vertical direction. The upward liquid velocity is sufficient to suspend the activated carbon so that the carbon does not have constant interparticle contact. At the top of the carbon there is a distinct interface between the carbon and the effluent water. The principal advantage of the fluidized bed is that waters with appreciable suspended solids content may be given adsorption treatment without clogging the bed, since the suspended solids pass through the bed and leave with the effluent.

Although the treatability of a particular groundwater by carbon and the relative capacity of different types of carbon for treatment may be estimated from adsorption isotherms, carbon performance, and design criteria are best determined by pilot column tests. Design-related information that can be obtained from pilot tests include:

1. Contact time
2. Bed depth
3. Pretreatment requirements
4. Breakthrough characteristics
5. Head loss characteristics
6. Carbon dosage in pounds of pollutants removed per pound of carbon

The design of an activated carbon adsorption column can be accomplished by using kinetic equations that require data obtained from the development of breakthrough curves.

8.4.3   Biological Treatment

In biological treatment of contaminated groundwater, the objective is to remove or reduce the concentration of organic and inorganic compounds. Because many of the compounds that may be present in contaminated groundwater can be toxic to microorganisms, pretreatment of the groundwater may be required. When a groundwater containing organic compounds is contacted with microorganisms, the organic material is removed by the microorganisms through metabolic processes. The organic compounds may be used by the micro-organisms to form new cellular material or to produce energy that is required by the microorganisms for their life systems.

Heterotrophic microorganisms are the most common group of microorganisms providing the metabolic process for removing organic compounds from contaminated groundwater. Heterotrophs use the same substances (organic compounds) as sources of both carbon and energy. A portion of the organic material is oxidized to provide energy, while the remaining portion is used as building blocks for cellular synthesis. Three general methods exist by which heterotrophic microorganisms can obtain energy. These are fermentation, aerobic respiration, and anaerobic respiration.

In the case of fermentation, the carbon and energy source is broken down by a series of enzyme-mediated reactions that do not involve an electron transport chain. In aerobic respiration, the carbon and energy source is broken down by a series of enzyme-mediated reactions in which oxygen serves as an external electron acceptor. In anaerobic respiration, the carbon and energy source is broken down by a series of enzyme-mediated reactions in which sulfates, nitrates, and carbon dioxide serve as the external electron acceptors. The three processes of obtaining energy forms the basis for the various biological wastewater treatment processes.

Biological treatment processes are typically divided into two categories: suspended growth systems and fixed-film systems. Suspended growth systems are more commonly referred to as activated sludge processes, of which several variations and modifications exist. The basic system consists of a large basin into which the contaminated water is introduced, and air or oxygen is introduced by either diffused aeration or mechanical aeration devices. The microorganisms are present in the aeration basin as suspended material. After the microorganisms remove the organic material from the contaminated water, they must be separated from the liquid stream. This is accomplished by gravity settling. After separating the biomass from the liquid, the biomass increase resulting from synthesis is wasted and the remainder is returned to the aeration tank. Thus, a relatively constant mass of microorganisms is maintained in the system. The performance of the process depends on the recycle of sufficient biomass. If biomass separation and concentration fails, the entire process fails. The process requires the skills of well-trained operators.

Fixed-film biological processes differ from suspended growth systems in that microorganisms attach themselves to a medium that provides an inert support. Biological towers (trickling filters) and rotating biological contactors are the most common forms of fixed-film processes. Biological towers are a modification of the trickling filter process. The media, which is normally comprised of polyvinyl chloride (PVC), polyethylene, polystyrene, or redwood is stacked into towers that typically reach 16–20 ft in height. The contaminated water is sprayed across the top, and, as it moves downward, air is pulled upward through the tower. A slime layer of microorganisms forms on the media and removes the organic contaminants as the water flows over the slime layer.

A rotating biological contactor (RBC) consists of a series of rotating discs, connected by a shaft, set in a basin or trough. The contaminated water passes through the basin where the microorganisms, attached to the discs, metabolize the organics present in the water. Approximately 40% of the disc surface area is submerged. This allows the slime layer to alternately come into contact with the contaminated water and the air where oxygen is provided to the microorganisms.

Removal efficiencies are generally the same for fixed-film and suspended growth processes. However, fixed-film processes have the potential to be lower in cost, due to the absence of aeration equipment, and are easier to operate. Both systems may be operated under anaerobic conditions, which may offer advantages for certain contaminated waters.

The addition of powdered activated carbon (PAC) to the activated sludge process has received considerable attention, particularly with respect to the removal of specific organics. The applicability of activated carbon in removing specific substrates depends on the molecular weight, solubility, polarity, location of functional groups, and overall molecular configuration. Investigations of PAC systems have centered around process enhancement factors. These include:

1. Enhancement of a system to receive shock loadings or temperature changes
2. Improved nonbiodegradable organics removal
3. Improved removal of specific organics
4. Improved color removal
5. Enhanced resistance to biologically toxic or bacteriostatic substances
6. Improved hydraulic capacity of existing facilities
7. Improved nitrification
8. Suppressed foaming in the aeration system
9. Improved sludge handling
10. Reducing sludge bulking

In recent years research has been primarily directed toward removal of priority pollutant organics, removal of other residual organic compounds, enhancement of nitrification, and improvement in sludge settlability.

8.5   TREATMENT TRAINS

Due to the complex composition of most groundwaters, no one unit operation is capable of removing all of the contaminants present. It may be necessary to combine several unit operations into one treatment process to effectively remove the contaminants required. In order to simplify and make visible the selection of the applicable treatment trains, Table 8.1 is presented and shows a number of unit operations and the waste types for which they are effective.

8.6   COST COMPARISON — A CASE STUDY

At one refinery, the following groundwater flow and water quality conditions were determined to be applicable during hydrocarbon recovery operations:

Table 8.1 Summary of Suitability of Treatment Procedures

Remediation Alternatives	Volatile Organics	Nonvolatile Organics	Inorganics
Air stripping	Suitable for most cases	Not suitable	Not suitable
High-temperature air stripping	Effective removal technique	May be suitable	Not suitable
Steam stripping	Effective concentrated technique	May be suitable	Not suitable
Carbon adsorption	Effective removal technique	Effective removal technique	Not suitable
Biological	Effective removal technique	Effective removal technique	Not suitable metals toxic
pH adjustment precipitation	Not applicable	Not applicable	Effective removal technology
Membrane processes	May not be applicable	Effective removal technique	Effective removal tecnique
Electrodialysis	Not applicable	Not applicable	Inefficient operation/ inadequate removal
Ion exchange	Not applicable	Not applicable	Inappropriate technology — difficult operation

Flow	500 gpm
BTX	20 mg/1 (total)
Iron	5–10 mg/1
Manganese	<1.0 mg/1

These water quality conditions are fairly typical for coproduced groundwater during hydrocarbon recovery operations at refineries. The principal treatment processes evaluated consisted of primary treatment for inorganics removal followed by air stripping and activated carbon adsorption. One alternative, also evaluated, consisted of primary treatment for inorganics removal followed by use as cooling tower make-up water and then biological treatment in existing refinery processes. However, this alternative may not be possible in most refineries due to the limited hydraulic capacities in their biological treatment systems.

8.6.1   Alternative 1

Alternative 1 consists of preliminary treatment for heavy metals removal with the primary concern being iron removal (Figure 8.3). The levels of iron observed in the groundwater at this site would be very detrimental to the downstream treatment processes. This pretreated water would then be used for cooling tower make-up water followed by biological treatment. This approach would be the easiest and cheapest alternative. This combined process should provide effective removal of BTX.

Two advantages of this approach would be the minimized capital and operating costs, along with minimal space requirements. The best process (for iron removal) would probably be chemical precipitation with lime through a rapid mix, flocculation, and dissolved air flotation step (DAF). DAF size requirements would be in the range of a 300- to 350-ft^2 unit. Total costs for operation and maintenance of the preliminary treatment facility using an interest rate of 10% over 10 years are estimated at $0.44/1000 gal. A summary of the capital and operating cost estimates are presented in Table 8.2.

8.6.2   Alternative 2

Alternative 2 consists of preliminary treatment followed by dual-media pressure filtration, and two-stage air stripping (Figure 8.4). The preliminary treatment step for iron removal would be exactly the same as specified under alternative 1. The filters would be recommended to remove suspended matter and particulate iron prior to the air strippers. The required filtration capacity could be provided with either a duplex system of two 60-in. diameter filters or a triplex system of three 42-in. diameter filters.

A two-stage air stripping system would be required to get the BTX levels down to nondetectable levels. The first stage would be designed to achieve effluent levels of 100–200 ppb, with the latter stage getting down to the level of detection. Preliminary sizing indicated four 4-ft diameter columns with two columns in the first stage and two columns in the second stage. The off-air emission from the first stage would require treatment to prevent atmospheric discharges of BTX. The emission control system was costed out for activated carbon treatment of the air emissions with steam regeneration; however, both the capital and operating costs could be reduced significantly if there were incinerator(s) on site that could use the off air for combustion air. The total cost with an emission control system was estimated to be $1.28 per 1,000 gallons.

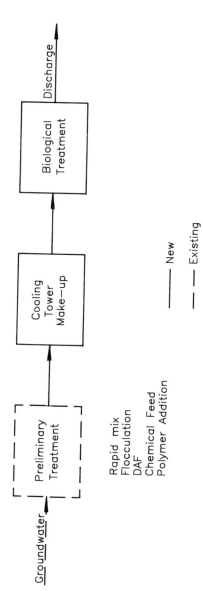

Figure 8.3. Preliminary treatment, cooling towers, and biological treatment (alternative 1).

Table 8.2   Summary of Total Annual Estimated Cost for Groundwater Treatment Alternatives

	Alt 1	Alt 2	Alt 3 Without Regeneration[a]	Alt 3 With Regeneration[b]
Capital cost	$216,000	$800,000	$634,000	$634,000
Annual projected cost at 10% interest, 10 years	$ 35,000	$130,000	$103,000	$103,000
Annual operation and maintenance cost	$ 81,500	$205,500	$638,000	$372,500
Total annual estimated cost	$116,500	$335,500	$741,000	$475,500
Actual water treated cost ($1,000 gal)	$ 0.44	$ 1.28	$ 2.82	$ 1.81

[a] Costs do not include sludge dewatering and disposal costs.
[b] Existing carbon regeneration furnace on-site.

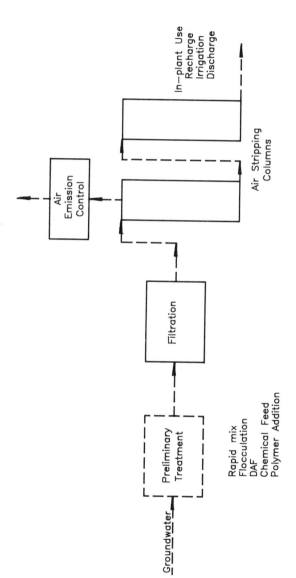

Figure 8.4. Preliminary treatment, filtration, and two-stage air stripping (alternative 2).

8.6.3   Alternative 3

Alternative 3 consists of preliminary treatment followed by filtration and activated carbon adsorption (Figure 8.5). The preliminary treatment step and filtration requirements and costs would be the same as specified for alternative 2 (air stripping).

A two-stage activated carbon system would also be required to get the BTX levels down to nondetectable levels. Two 10-ft diameter columns should provide the required treatment capacity. Each column would be charged with 20,000 lbs of carbon. An estimated carbon usage rate of 2 lbs/1,000 gal of water treated was used for cost estimating. Costs were estimated both with and without on-site regeneration of the carbon. As can be observed in Table 8.2, it would be cheaper to regenerate on-site. The difference in costs is estimated at $1.80 vs. $2.82 per 1,000 gal of water treated. Costs for carbon regeneration were based on using an existing multiple hearth carbon regeneration furnace on-site at the refinery. A total breakdown of costs (capital and O&M) is presented in Table 8.2. These costs do not include sludge dewatering and disposal facilities.

8.7   DISPOSAL OPTIONS

Typically at LNAPL recovery sites (i.e., refineries and bulk terminals), some quantity of groundwater is coproduced. A major concern arises from this process in that the coproduced water must be treated, disposed of, or both. An evaluation process then follows on how to handle the coproduced water. A number of factors that control the ultimate fate of the water include: average volume produced on a regular basis, level of contamination, and site-specific physical constraints.

It is one thing to design, test, and maintain a recovery system that effectively recovers LNAPL, but the handling of the coproduced groundwater is another matter. The effectiveness of a recovery system can suffer if the disposal capacity is insufficient or treatment costs are high. Additionally, a cost-effective, low operation and maintenance recovery system can quickly be turned into a high cost system if the coproduced water fate is not coordinated with the recovery system. A key factor will be the level of treatment required, if any, prior to disposal. The three following alternatives are explored: surface discharge; refinery reuse; and reinjection.

8.7.1   Surface Discharge

Discharge of the coproduced water without treatment is dependent upon the level of contamination, the volume produced, and limitations set forth by existing discharge permits. If the concentration levels are low enough and a

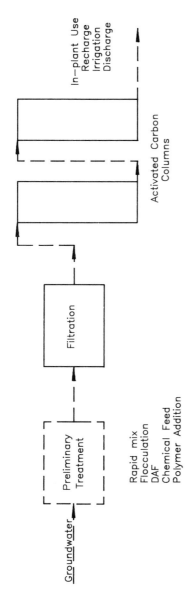

Figure 8.5. Preliminary treatment, filtration, and granular activated carbon adsorption (alternative 3).

discharge permit (with regular sampling) is in place, generally only a volume limit adjustment to the permit is required; however, discharge permits are difficult to obtain in certain areas. In terms of operation and maintenance (O&M) costs, surface discharge is a cost-effective option. At most large facilities, refinery and bulk terminals, retention basins (for stormwater runoff) and/or oil/water separators are available. In-line placement of one or both of these features prior to surface discharge may be desirable depending upon volume and concentration limits imposed by the permit. The oil/water separator will be utilized as a safety precaution in the event that the LNAPL is discharged. The retention ponds can accept cascading coproduced water, which may act as a passive air stripping method. Concentration levels could be reduced through off-gassing.

Surface discharge without treatment will ultimately depend on the total mass of contaminants discharged and the acceptable mass loading rate for the particular discharge point. Therefore, recovery programs producing groundwater at low volumes with high BTX concentrations or high volumes with moderate BTX concentrations will likely require some level of pretreatment prior to direct discharge.

8.7.2   Refinery Reuse

Refinery reuse is a valid disposal option. Again, depending upon the level of contamination and the ability of the recovery system to provide a constant volume, this option can be a desirable one if the facility is presently purchasing large quantities of import water from local water utilities. With this option, costs for disposal design, construction, and O&M are eliminated. In addition, a reduction in the refinery's water bill can be realized. Problems generally encountered with this option include the natural quality of the groundwater, including iron and manganese content, and alkalinity. Typically, minor or passive treatment methods, such as oil/water separators and retention pond settling, can eliminate more costly treatment methods depending upon the use of the water. One possible option is to use the water for cooling tower make-up water. Depending on the volume, this make-up water could then be routed through the existing wastewater treatment facilities prior to direct discharge.

8.7.3   Reinjection

Reinjection of coproduced groundwater through the use of wells is commonly used to return the water to the same aquifer and to set up hydraulic barriers in an effort to contain the plume. Injection wells are commonly used in conjunction with withdrawal systems to enhance the recovery of hydrocarbons. Injecting water at appropriate locations will create a pressure ridge to effectively increase the hydraulic gradient towards the withdrawal point. Normally, the water pumped from the recovery wells is used as the injection water and is injected without treatment. This method provides an economical way of handling the produced water, as well as being beneficial to the recovery effort.

In the United States, the Environmental Protection Agency (EPA) is currently considering a ban on this type of injection by reclassifying these injection wells as Class IV wells or wells that inject hazardous substances into or above a potential underground source of drinking water. Class IV injection wells are banned by law in the United States. There is technical justification to argue against this potential ban at hydrocarbon recovery sites. As long as free and residual hydrocarbon are present above the water table in the area of reinjection, there will be continuous contamination by dissolved fractions. The argument for reinjection without treatment would be that the injection water is the same quality of water already present within the aquifer, as long as it is reinjected into the same geologic horizon from which it was originally pumped.

The issue of dissolved fractions — in terms of aquifer restoration objectives — would be more appropriately addressed at a phase of the remediation program after the source of the dissolved contamination is controlled. Of course, containment of the dissolved plume remains a priority throughout the program. For large-scale recovery programs, requiring treatment prior to reinjection would place an excessive economical burden on the overall remediation effort without technical justification, at least during the course of free hydrocarbon recovery.

Reinjection wells are usually constructed of either 2-, 4-, or 6-in. PVC. The screen portion of the well usually has slotted openings. Oversized screen openings and larger gravel pack are normally designed for reinjection wells as opposed to withdrawal well design. The reason for the overdesign is to gain the most open area possible. Also, groundwater is being injected into the well, placing a positive pressure on the formation. Therefore, silting of the well from this overdesign is not of major concern. However, gross overdesign should be avoided, because eventually the well will have to be redeveloped and the most effective way to develop the reinjection well is to reverse the flow by pumping or air lifting.

Groundwater quality at most refinery and terminal facilities is not of pristine quality. Elevated concentrations of iron, manganese, BTX, and petroleum hydrocarbons exist. Upon exposure to the atmosphere or air mixing, changes begin to occur in the make-up of the coproduced groundwater. Iron will precipitate out and form a scale on the reinjection well openings. This will reduce the efficiency of the reinjection well quickly. Iron bacteria will also form when three important components are present. These are iron, air, and dissolved petroleum hydrocarbon. If a reinjection well is allowed to degrade to such a point whereby significant amounts of iron bacteria have formed in the screen openings of the well, the gravel pack, and the surrounding formation, redevelopment of the well may not be feasible. Reinjection systems should, therefore, if possible, employ the logic of a closed system (i.e., eliminate exposure of the water to the atmosphere). To do this, the system must be constructed with a closed pipe from the discharge point to the reinjection well. At the reinjection well, a drop pipe is installed usually with a flow restrictor at the bottom to reduce the potential for cascading. The bottom of the drop tube is terminated at or near the bottom of the well, beneath the static fluid level, also to aid in the prevention of cascading water.

An important factor to keep in mind when designing reinjection wells or systems is that eventually they will have to be redeveloped. Reinjection wells and the redevelopment of these wells unfortunately follow the law of diminishing returns. Constant reinjection of coproduced groundwater without regular development results in irreversible clogging and ultimately the abandonment of the well.

The location and spacing of reinjection wells is a design area that is sometimes overlooked. Most of the design effort is placed in constructing the most efficient and strategically located recovery well. Granted, the recovery well design is extremely important, but the location and spacing of reinjection wells can be critical in the total system performance. Proper spacing and location of reinjection wells can be advantageous in the overall scheme. Strategic placement of the injection wells upgradient or downgradient in relationship to the free-phase hydrocarbon plume and the distances are all realistic considerations.

Placing the reinjection wells in a relative upgradient location, in relationship to the free-phase hydrocarbon plume, forms a groundwater mound that creates an increased gradient in the direction of the recovery well. This, of course, is a desirable effect where the ultimate goal is to recover the largest amount of free-phase hydrocarbon as possible in a shorter time frame.

Locating reinjection wells downgradient of the plume can be advantageous as well. In this instance, a line of downgradient reinjection wells creates the same groundwater high or mound. This mound in essence forms a groundwater wall or barrier. The barrier serves two purposes: first, the barrier can prevent off-site migration; and second, the barrier blocks the migrating free-phase hydrocarbon, causing it to accumulate in this area for recovery. Obviously, the location of recovery wells and reinjection wells is closely associated.

The location of the reinjection wells with respect to the recovery well is an important consideration. Locating the wells too closely can create counterproductive effects. Locating the reinjection wells too far away from the recovery well may significantly reduce the desired effect or eliminate the benefits altogether. If the latter is the case, "dead spots" may result and may cause short circuiting of the free-phase hydrocarbon plume. Once this happens, off-site migration of the plume is possible. In fact, reinjection wells placed and spaced improperly can, in some instances, accelerate off-site migration.

Reinjection wells are not without disadvantages. One disadvantage referred to above is the possibility of accelerating off-site migration or dispersion of the plume. Another disadvantage is the additional costs for operation and maintenance (i.e., redevelopment). Redevelopment costs can be kept to a minimum by initial design considerations, but, as the time frame between redevelopment decreases, redevelopment costs increase. Eventually the cost of constructing a new reinjection well becomes the only cost-effective alternative.

In summary, when undertaking a project such as the recovery of LNAPL, treatment of the coproduced water, prior to reinjection, may not be beneficial or technically necessary. A large percentage of the spilled or leaked petroleum hydrocarbon (40–60%) will be retained in the unsaturated zone as residual saturation. This residual hydrocarbon cannot be recovered by conventional withdrawal techniques. Without removing this continual source of contamination to the groundwater system, dissolved contamination will continue. Therefore, in most cases, it may be pointless and extremely costly to treat the coproduced groundwater prior to reinjection while the free- and residual-phase hydrocarbon contamination exists.

REFERENCES

1. Bellandi, R., 1988, Hazardous Waste Site Remediation, *The Engineer's Perspective*, O'Brien & Gere Engineers, Inc.: Van Nostrand Reinhold, New York, New York, 422 p.
2. Dougherty, P. J. and Paczkowski, M. T., 1988, A Technique to Minimize Contaminated Fluid Produced during Well Development: In Proceedings of the Association of Ground Water Scientists and Engineers and USEPA Second National Outdoor Action Conference on Aquifer Restoration, Ground Water Monitoring, and Geophysical Methods, May, 1988, Vol. I, p. 303-317
3. Driscoll, F. G., 1986, *Groundwater and Wells*: Johnson Division, St. Paul, MN, 2nd ed., 1089 p.
4. Ford, D. L., 1980, Technology for Removal of Hydrocarbon from Surface and Groundwater Resources: Engineering Science, Inc., and the University of Texas, Austin, Texas.
5. Fryberger, J. S. and Shepard, D. C., 1987, Reinjection of Water at Hydrocarbon Recovery Sites: In Proceedings of the International Symposium on Class V Injection Well Technology, USEPA and UIPC, September, 1987, Washington, D.C.
6. Kaufmann, H. G., 1982, Granular Carbon Treatment of Contaminated Supplies: In Proceedings of the National Water Well Association Second National Symposium on Aquifer Restoration and Ground Water Monitoring, May, 1982, p. 94-104.

7. Knox, R. C., Canter, L. W., Kincannon, D.F., Stover, E.L., and Ward, C.H., 1986, *Aquifer Restoration - State of the Art*: Noyes Publications, Park Ridge, NJ, 490 p.

8. Lamarre, B. L., McGarry, F. J., and Stover, E. L., 1983, Design, Operation and Results of a Pilot Plant for Removal of Contaminants from Ground Water: In Proceedings of the National Water Well Association Third National Symposium on Aquifer Restoration and Ground Water Monitoring, May, 1983, p. 113-122.

9. Olsthoorn, I. T. N., 1982, The Clogging of Recharge Wells, Main Subjects, The Netherlands Waterworks Testing and Research Institute, KIWA.

10. Stover, E. L., 1982, Removal of Volatile Organics from Contaminated Ground Water: In Proceedings of the National Water Well Association Second National Symposium on Aquifer Restoration and Ground Water Monitoring, May, 1982, p. 77-84.

11. Stover, E. L. and Kincannon, D. F., 1982, Treatability Studies for Aquifer Restoration: Presented at the American Water Works Association 1982 Joint Annual Conference, Southwest and Texas Sections, Oklahoma City, OK.

12. Stover, E. L. and Kincannon, D. F., 1983, Contaminated Groundwater Treatability - A Case Study: *J. Am. Water Works Assoc.*, Vol. 75, No. 6, p. 292-298.

13. Stover, E. L., 1983, Priority Pollutant Treatability Studies: Presented at Chemical Manufacturers Association Seminar on Biological Treatment, Priority Pollutants and BATEA, Washington, D.C.

14. Stover, E. L., Gomathinayagam, G., and Gonzalez, R., 1985, Design Guidelines for Biodegradation of Specific Organic Pollutants in Petroleum Refinery Wastewaters: Presented at the Energy-Sources Technology Conference and Exhibition, Dallas, Texas.

15. Stover, E. L., Fazel, A., and Kincannon, D. F., 1985, Powdered Activated Carbon and Ozone Assisted Activated Sludge Treatment for Removal of Toxic Organic Compounds: Ozone: Science and Engineering, Vol. 7, No. 3, p. 191-203.

16. Stover, E. L., Gates, M. M., and Gonzalez, R., 1986, Treatment and Removal of Dissolved Organics and Inorganics in a Contaminated Ground Water - A Case Study: In Proceedings of the National Water Well Association and American Petroleum Institute Conference on Petroleum Hydrocarbons and Organic Chemicals in Ground Water - Prevention, Detection and Restoration, Houston, TX, p. 689-708.

17. Stover, E. L., 1989, Coproduced Ground Water Treatment and Disposal Options during Hydrocarbon Recovery Operations: *Ground Water Monitoring Review*, Winter Issue, Vol. 9, No. 1, p. 75-82.

18. Wilson, J. T., Leach, L. E., Michalowski, J., Vandegrift, S., and Callaway, R., 1989, *In Situ* Bioremediation of Spills from Underground Storage Tanks: New Approaches for Site Characterization, Project Design, and Evaluation of Performance: United States Environmental Protection Agency Report No. EPA/60012-89/042, July, 1989, 56 p.

9 SOIL REMEDIATION ALTERNATIVES

"The microscopic organisms are very inferior in individual energy to lions and elephants, but in their united influence they are far more important than all of these animals" (Ehrenburg, 1862)

9.1   INTRODUCTION

In virtually every situation where the source of a petroleum spill occurs at or near the land surface, some quantity of product is retained in the soil as residual saturation within the unsaturated zone above the water table. Most unsaturated (with water) soils are capable of holding petroleum product in a quantity equivalent to approximately 30% of their water-holding capacity (the amount of water that a soil can hold at saturation). In most cases, however, by the time remediation of the soil is implemented, the residual retained product is less than maximum.

Removal (or destruction) of the spilled product from the unsaturated zone can be considered and designed similarly to that of an industrial process procedure. A bulk of material must be treated to eliminate an unwanted contaminant. The method used, regardless of the option chosen must meet certain criteria:

- Eliminate the problem caused by the product
- Accomplish the task within a reasonable time frame
- Be economically acceptable

Definition of *eliminate, time frame,* and *economically acceptable* are terms to be resolved between the designer, client, and regulatory agency. After consideration of these basic criteria, many procedures can be made to work, either independently, sequentially, or in combination. Some of the more common processes that have been used (or experimented with) to remediate petroleum-contaminated soils in the unsaturated zone are discussed below. Conventional processes are available to assist the designer in screening possible solutions to specific site problems. A summary of each option, including a brief discussion of process, estimated costs, practical constraints, and miscellaneous remarks, is presented in Table 9.1. Also presented in Table 9.1 are references regarding methodology techniques and pertinent case studies per option. Since an in-depth discussion of remediation of petroleum-contaminated soils is beyond the scope of this book, more comprehensive discussion can be found in Kostecki and Calabrese (1989), Calabrese and Kostecki (1989), and Dragun (1988).

9.2   EXCAVATION, TREATMENT, AND DISPOSAL

Physical removal of the soil is an effective, but also the most expensive procedure per cubic yard (excluding incineration) to remove contaminants from the source and the affected area. Conventional construction equipment can be used. Excavation is commonly practiced during the removal of existing underground storage tanks. Often the necessary equipment is already on site or readily available. Once the soil is removed, several options are available for disposal.

9.2.1   Direct Transport to a Licensed Landfill

If the soil meets criteria acceptable to the landfill, which may be difficult at times, the soil can be disposed of there. Normally, soils are considered acceptable as long as the maximum concentration of volatile fuel is below 1000 ppm, no flammable vapors are present, and no other constituent that may be considered hazardous is present.

9.2.2   Excavation and Aeration and Disposal at a Landfill

If the concentration of fuel is initially too high to be accepted at a landfill, it may sometimes be spread as a thin layer, usually less than 1 ft in thickness, and tilled to aerate it with conventional agriculture equipment. After sufficient volatile components have evaporated to reduce the concentration below acceptable levels, the soils are then transported to a landfill.

9.2.3   Excavation, Landfarming, and Replacement

Soils that are highly contaminated with semivolatile or nonvolatile fuels may be treated on-site prior to either reuse or for off-site disposal. If climatic conditions are suitable, sufficient space is available, air emission regulations are flexible, and time is not a major factor, it may be possible to "landfarm" the soil. This procedure involves spreading the soil in a thin layer, applying moisture if necessary, and allowing a combination of aeration and biodegradation to take place for removal of the hydrocarbon constituents. When less volatile products are involved, it may be necessary to enhance the bioactivity with added bacteria or to add nutrients to stimulate the clean-up process. Continued tilling and moisture addition are necessary for optimum performance. After testing demonstrates that the contaminants are significantly reduced, the soil may be recompacted in the original excavation.

9.2.4   Mechanically Enhanced Volatilization

Where on-site treatment is necessary, but landfarming is not possible, volatile hydrocarbons and fuel products may be removed by the use of a mechanical aerator. Several variations of this process are available, including rotary mechanical aeration, low-temperature thermal air stripping, and a pneumatic conveyor system. These processes are discussed below.

Rotary mechanical aeration is accomplished by the use of a pugmill or a rotary drum, which mixes the soil in a continual flow of clean air. Volatile components are transferred to the air for treatment elsewhere (or discharge). Because the unit is closed, the vapors are controlled and do not restrict other activities at the site.

Low-temperature thermal air stripping is similar to mechanical aeration, except that the soil is heated during or prior to treatment. Usually this process involves the use of a screw auger or rotary drum, which continually mixes the soil in the presence of a flow of clean air. The contaminated air is often passed through an afterburner or catalytic converter prior to discharge.

A pneumatic conveyor system can also effectively remove volatile products. This process involves using high-velocity air to convey pulverized air through a long duct. Volatile product is transferred from the soil to the air, while clean soil is retrieved at the end of the duct from a cyclone. Mass transfer of fuel can be improved by the use of preheated air.

Of the three processes discussed above, low-temperature thermal stripping has shown the most promise for practical cost-effective soil restoration.

Table 9.1   Soil Remediation Options

Option	Process	Estimated Costs[a]	Practical Constraints	Remarks
Excavation and disposal as hazardous waste	Excavate and haul to class I landfill; emplace and compact clean fill	$300/yd^3	Cradle-to-grave liability as waste generator	High cost
Excavation and disposal as solid waste (nonhazardous)	Excavate and haul to class III landfill; backfill with clean fill	$ 60/yd^3	Location of a suitable landfill	Economical on small projects
Excavation, aeration and disposal off site	Excavate and spread on site; turn repeatedly to aerate, haul to clean fill disposal site; emplace and compact new clean fill	—	Emission considerations	Technically feasible; permitting very difficult under current legislation; requires numerous analytical tests
Excavation, landfarming and replacement	Excavate and spread on-site; aerate and add nutrients and water; re-emplace and compact	$ 50/yd^3	Emission considerations; leaves excavation open during treatment	Technically feasible, permitting may be difficult; requires numerous analytical tests
Mechanically enhanced volatilization	Excavate, pass through crusher, aerator, and re-emplace	$250/yd^3	Requires dust control and vapor treatment	High cost, but suitable for specific locations

In situ venting (vacuum extraction)	Investigate extent of contamination and soil conditions; design and install venting system; permit system; operate system; re-investigate to monitor effectiveness	$ 20–50/yd^3	Fine-grained soils and low volatility of hydrocarbon in soils limit the effectiveness of this method	Not a technically viable option for sites with clayey soils; requires disposal of air filtration medium
In situ biodegradation or chemical degradation	Investigate extent of contamination and soil and groundwater conditions; conduct feasibility study; design and install pumping and injection system; permit system; operate system; reinvestigate to monitor effectiveness	$ 75/yd^3	Fine-grained soils limit ability to inject andpump fluids through soils. System could be engineered to be installed and operated around existing facilities; requires on-going operation and maintenance (O&M) and monitoring. Requires periodic soil sampling and final investigation	Overall effectiveness cannot be assured; pending results of pilot study; requires on-site monitoring

Table 9.1   Soil Remediation Options (continued)

Option	Process	Estimated Costs[a]	Practical Constraints	Remarks
Steam injection and stripping	Investigate extent of contamination and soil and groundwater conditions; conduct feasibility study; design and install steam injection and recovery system; permit system; operate system; monitor effectiveness on an on-going basis	$100–200/yd^3	Fine-grained soils limit ability to inject steam and recover fluids from soils	Overall effectiveness cannot be assured, pending pilot study results; high O&M cost
Asphalt incorporation	Excavate and transport	$125/ton	Soil must pass flash-test before acceptance	Very good option for low-volatile hydrocarbon affected soils
No action	No action		No risk to public health, safety and welfare. No risk to surface water or ground water; considered of beneficial use	Site-specific

Soil washing/extraction	Excavate, crush, mix with wash fluid, separate, replace, treat wash water	Limited to granular soils, wash fluid treatment may be difficult	Technically feasible, high cost, limited applications
In-place leaching	Construct infiltration and recovery systems, irrigate washing fluid, retrieve fluid, treat fluid	Limited to permeable soils, and higher solubility hydrocarbons	Often used in conjunction with bio-treatment practices; permit approval may be difficult
Above-ground leaching/ replacement	Excavate, crush, place over collector bed, flush with wash fluid, replace, treat fluid	Total washing fluid collection, temperature and odor control; requires fairly large open area	May be used in association with bio-treatment, often effective; permitting not as difficult

[a] Estimated costs reflect 1990 dollars.

9.3   SOIL VENTING

This *in situ* technique was originally designed to control vapor-phase hydro-carbon that had migrated into basements of buildings, storm sewers, and utility vaults. This technique has since been developed and used in recent years to remove volatile products from unsaturated soils. Soil venting involves drawing air through the vadose zone via vapor recovery wells or other extraction points. Volatile hydrocarbon occurring as residual saturation transfers to the air and is withdrawn through the extraction points. Vapor wells or extraction drains (analogous to French drains) are constructed in a similar manner to an ordinary monitor well or drain, except that they are completed in the vadose zone. Generally, vacuum pumps, blower fans, or both are used to draw air through the formation and out of the extraction points. The American Petroleum Institute sponsored a large pilot-scale study of soil venting in a moderately permeable formation. The venting geometry for this study was designed to maximize lateral flow and the migration of vapors in the soil. This pilot study demonstrated that venting was extremely effective in reducing and controlling the concentration of hydrocarbons in the soils over a large area of the vadose zone. The radial influence of the system extended beyond 110 ft using air flow rates of approximately 40 standard cubic feet per minute and resulted in the recovery of vapors equivalent to 5.86 gal/d of liquid hydrocarbon.

Some limitations are apparent to the effectiveness of soil venting. For example, spills involving less volatile hydrocarbons may not be a candidate for soil venting techniques without the injection of steam or hot air to volatilize those components. In areas of extremely low permeability, such as clay, it may not be possible to draw air effectively through the contaminated zone, even with multiple withdrawal points or drains.

9.4   SOIL WASHING/EXTRACTION

This generic title is used to describe a variety of processes that use a liquid leaching medium to extract contaminants from soil. The same principles apply whether the soil is treated in the subsurface or excavated for above-ground treatment. Normally water is used as the flushing medium, and often a surfactant is added to enhance performance. The effectiveness of this type of operation is dependent upon the capacity of the soil to retain the fuel product. Diesel fuel, kerosene, and gasoline are less tightly held than creosote or coal tar. Typical procedures for soil leaching are described in the following sections.

9.4.1   Leaching In-Place

Field practice of this operation has been performed as both vertical flushing using surface irrigation, which is recovered by wells or drain pipes at depth, and also horizontal flushing by the use of injection and recovery wells. The success of operation depends upon the nature of the product, the type of soil matrix, the presence of other organic matter in the soil (i.e., free organic content or biomass), and the temperature of operation.

Surfactants can be useful to reduce the surface tension between the flushing water and the product. However, the surfactant used should be either hydraulically fully recoverable or easily biodegradable. Several experiments using oil-field surfactants have been successfully used to remove petroleum products from laboratory samples of freshwater aquifer materials; however, the surfactants used were not biodegradable. The use of such materials in aquifer settings is not acceptable.

Some naturally occurring soil microbes develop enzymes that act as surfactants. This process appears to be their method of freeing the oil for their own consumption, however, it can also be useful for planned remediation. Biological laboratory studies can often identify the processes that can improve this action.

9.4.2   Leaching Above-Ground

Several mechanical systems have been developed to wash soil in above-ground facilities. These "washing machines" utilize pugmills, feed augers, or other mixing equipment to blend soil and washwater (or special solvent) to extract the petroleum product. Water (with surfactant) is the most commonly used flushing agent. Product remaining after washing is relatively immobile under normal environmental conditions. Wasted water and surfactant can be treated by conventional treatment means.

Special solvents that are occasionally used have the physical property that they are soluble in water at normal temperature (20° C), but are almost insoluble at higher temperatures. These solvents can be mixed with water to extract petroleum products from pulverized soils. After separation from the solids, the liquid is heated to separate into the water phase and the insoluble oil phase. The solvent is then separated from the product and recycled. The soil is dewatered by conventional means (such as vacuum filter, filterpress, or centrifuge) and used as fill material.

9.5   MICROBIAL DEGRADATION

Natural soils in the aerobic region almost always contain microorganisms that can use fuel-type hydrocarbons as an energy source. Soil bacteria, actinomyces, and other microbes can relatively easily be acclimated to utilize the compounds in their metabolic process. Petroleum product is reduced to biomass and carbon dioxide during this type of degradation.

The petroleum industry has utilized this process extensively for treatment of refinery wastes by "landfarming". This technique involves blending wastes into shallow surface soils in the presence of added nutrients (nitrogen, phosphorous, and potassium) and moisture, which allows microorganisms (naturally occurring or introduced) to degrade the wastes. Continued tilling (with agricultural implements) to ensure aeration and constant monitoring are important steps to optimize performance.

Residual hydrocarbons in the unsaturated zone resulting from spills are subject to the same processes. Properly engineered remediation activities can provide the necessary components to encourage timely degradation of hydrocarbons in this region. In general terms, the primary task is to either deliver the biomass and nutrients to the hydrocarbon or the reverse.

Degradation of hydrocarbon by biological means at any given site will be dependent upon

- Indigenous soil microbial population
- Hydrocarbon variety and concentration
- Soil structure
- Soil reaction
- Nutrient Availability
- Moisture content
- Oxygen availability
- Whether previous releases have occurred

Soil microorganisms reported to degrade hydrocarbons under favorable conditions include: *Pseudomonas, Flavobacterium, Achrombacter, Anthrobacter, Micrococcus,* and *Acinobacter.* Over 200 species of soil microbes can assimilate hydrocarbon substrate. Fertile soils usually contain 10 EE7 to 10 EE9 microbes/g of dry soil, of which 10 EE5 to 10 EE6 are hydrocarbon degraders (prior to addition of hydrocarbons). After hydrocarbons have been introduced, it is common to observe hydrocarbon degraders at 10 EE6 to 10 EE8/g of dry soil.

Sites that do not naturally contain hydrocarbon degraders or that are in low numbers may be augmented by the addition of acclimated organisms from outside sources. Several commercial companies specialize in supplying these bacteria.

The variety of hydrocarbon product and its concentration also have a great effect on the rate of bioactivity. Hydrocarbons with less than 10 carbon atoms tend to be relatively easy to biodegrade, as long as the concentration is not high enough to be toxic to the organisms. Benzene, xylene, and toluene are examples of gasoline components that are easily degraded. As the molecule size increases, the biodegradation rate tends to decrease at an almost disproportionate rate. Complex molecular structures, such as branched paraffins, olefins, or cyclic alkanes, are much more resistant to biodegradation.

Soil structure, which is the form of assembly of the soil particles, determines the ability of that soil to transmit air, water, and nutrients to the zone of bioactivity. Soils that contain significant percentages of silt and clay do not usually transmit these substances rapidly enough to encourage rapid bioactivity. More permeable soils, such as sand, are more conducive to rapid clean-up activities.

Soil reaction is defined as a function of pH. Optimum oil/sludge digestion rates have been observed to occur at pH 7.5–7.8. Under more acidic conditions, fungi predominate in numbers and slow the reactions. A mixed balance of bacteria and fungi (most efficient) develops under slightly alkaline conditions.

Another major controlling factor is the variety and balance of nutrients necessary for microbic activity to proceed. Nutrients required by these microbes are similar to those of other plants. Nitrogen and phosphorous are the most common additives, although other minor compounds are occasionally lacking in the natural setting. As an example, one study found that one liter of gasoline required 44 g of nitrogen, 22 g of phosphorous, and 760 g of oxygen. Physical conditions, product quality, and biomass are unique to every site. Laboratory testing is a prerequisite to the successful project.

The majority of organisms that degrade fuel products function best near 20°C, although significant activity has been noted between 4°C and 40°C. Fortunately these temperatures are common in the subsurface for at least part of the year in most parts of the inhabited world.

Moisture availability is a very important factor in microbial action. The quantity of water present will influence the species composition at the site. In general, optimum activity occurs when the soil moisture is 50–80% of saturation (moisture holding capacity). When moisture content falls below 10%, bioactivity becomes marginal.

Oxygen availability controls the rate at which aerobic organisms can function. (Anaerobic digestion can occur, but the rate is much slower and has less complete digestion). Where pore spaces are filled with liquid, either petroleum or water, the supply of oxygen is greatly reduced compared to that of air. Lack of oxygen is usually the major limiting factor for *in situ* biodegradation. Groundwater may contain 1-4 ppm oxygen, while air contains much more. One liter of pore space containing 4 ppm oxygen contains approximately 4 mg oxygen. One liter of air containing 20% oxygen contains 256 mg oxygen. Bioactivity in unsaturated soils, therefore, can be much faster above than below the water table when an adequate air supply is provided, and oxygen and nutrient levels are controlled.

When biotreatment is planned to restore the unsaturated zone, design considerations must include the same dependent topics discussed earlier in this chapter. A preliminary (but detailed) soil study is necessary to describe the physical setting, including the soil characteristics, product type and distribution, depth to water table, degree of saturation, and initial biological activity. Also, a biological feasibility evaluation will determine the nutrient and oxygen supply requirements for optimal operation. The resulting project plan will include mechanical provisions to supply water and nutrients, to recover possible released hydrocarbons, and to ensure a continued supply of oxygen.

A representative project is discussed in Chapter 11. In that case, the nutrients and oxygen are added in the irrigation water, which is distributed sporadically over the treatment area to maintain a healthy biological environment (without soil saturation). Wells set below the water table maintain a drawdown cone under the site to retrieve any excess nutrients or freed product. Continued monitoring of recycled water and regular soil coring is used to document the restoration progress.

9.6   STEAM INJECTION AND HEATING

Injection of steam into the subsurface has been proposed as a method of enhancing volatilization or mobility of petroleum hydrocarbon in unsaturated soils. Recovered vapor products would be condensed, captured by carbon filters for recycling, or destroyed by a catalytic convertor.

This process offers promise, particularly at sites that have limited physical access, as well as volatile product and a large quantity of steam available. Control of the distribution of the steam and vapors escaping to the surface may be difficult, resulting in a difficult safety situation. Economic considerations must be evaluated fully, because the cost to heat a large mass of soil can be significant.

9.7   ASPHALT INCORPORATION

Another alternative to landfilling or on-site treatment of petroleum hydrocarbon is to recycle it into useful products such as asphalt paving material. In this process, sandy soils are preferred over clay soil, because a high percentage of clay is not a desirable component in asphalt paving materials.

After testing for petroleum hydrocarbon content (and presence of possible hazardous substances), the soil is delivered to the batch plant, where it is crushed and sieved through screens to remove wood, metal, or other undesirable debris. Next, the soil is passed through a gas- or oil-fired rotary kiln, where it is heated to approximately 350°F, which evaporates all the water and burns the petroleum hydrocarbon components. Thereafter, the soil is blended with other aggregates and asphalt, and delivered to the paving location.

Recycling by this procedure removes the contaminants from the original site, destroys undesirable materials, and provides a valuable construction material. Reasonable safety cautions during the initial hauling and storage, as well as adequate quality testing, are the only major technical difficulties related to asphalt batching. Local regulatory agencies may have administrative controls and should be contacted prior to using this process.

REFERENCES

1. Alexander, Martin, 1961, *Introduction to Soil Microbiology:* John Wiley and Sons, Inc., New York, New York.
2. Alexander, Martin and Lustigman, 1966, Effect of Chemical Structure on Microbial Degradation of Substituted Benzenes: *J. Agric. Food Chem.,* Vol. 14, No. 4, p. 410-413.
3. American Petroleum Institute (API), 1985, Detection of Hydrocarbons in Groundwater by Analysis of Shallow Soil Gas/Vapor: *American Petroleum Institute,* Publication, No. 4394.
4. American Petroleum Institute (API), 1985, Subsurface Venting of Hydrocarbon Vapors from an Underground Aquifer: *American Petroleum Institute Publication,* No. 4410.
5. Anastos, G., Corbin, M. H., and Coia, M. F., 1986, *In Situ* Air Stripping: A New Technique for Removing Volatile Organic Contaminants from Soils: In Superfund '86.
6. Baehr, A. L. and Hoag, G. E., 1986, A Modeling and Experimental Investigation of Venting Gasoline from Contaminated Soils. University of Massachusetts.
7. Bennedsen, M. D., 1987, Vacuum VOC's from Soil: *Poll. Eng.,* Vol. 19, No. 1, p. 66-68.
8. Bennedsen, M. D., Scott, J. P., and Hartley, J. D., 1987, Use of Vapor Extraction Systems for In Situ Removal of Volatile Organic Compounds from Soil: In Proceedings of the Hazardous Materials Control Research Institute National Conference on Hazardous Waste and Hazardous Materials, March, 1987, Washington, D.C., p. 92-95.
9. Bohn, H. L., 1977, Soil Treatment of Organic Waste Gases: *In Soils for Management of Organic Wastes and Waste Waters,* American Society of Agronomy, Madison, WI.
10. Borden, R. C. and Chih-Ming, K., 1989, Water Flushing of Trapped Residual Hydrocarbon: Mathematical Model Development and Laboratory Validation: In Proceedings of the National Water Well Association Conference on Petroleum Hydrocarbons and Organic Chemicals in Groundwater: Prevention, Detection and Restoration, November, 1989, p. 473-486.
11. Brookman, G. T., Flanagan, M., and Kebe, J. O., 1985, Literature Survey: Hydrocarbon Solubilities and Attenuation Mechanics: American Petroleum Institute of Health and Environmental Sciences Department, No. 4414, 101 p.

12. Brookman, G. T., Flanagan, M., and Kebe, J. O., 1985, Literature Survey: Unassisted Natural Mechanisms to Reduce Concentrations of Soluble Gasoline Components: American Petroleum Institute of Health and Environmental Sciences Department, No. 4415, August, 1985, 73 p.

13. Calabrese, E. J. and Kostecki, P. T., 1989, *Petroleum Contaminated Soils, Vol. 2:* Lewis Publishers, Chelsea, Michigan, 515 p.

14. Camp, Dresser and McKee, Inc., 1986, Superfund Treatment Technologies: A Vendor Inventory. EPA 540/2-86-004.

15. Castle, C., Bruck, J., Sappington, D., and Erbaugh, M., 1985, Research and Development of a Soil Washing System for Use at Superfund Sites: In Proceedings of the United States Environmental Protection Agency Sixth National Conference on Management of Uncontrolled Hazardous Waste Sites, Washington, D.C., p. 452-455.

16. Connor, J. R., 1988, Case Study of Soil Venting: *Poll. Eng.*, Vol. 20, No. 7, p. 74-78.

17. Crow, W. L., Anderson, E. P., and Minugh, E. M., 1987, Subsurface Venting of Vapors Emanating from Hydrocarbon Product on Ground Water: *Ground Water Monitoring Review*, Vol. 7, p. 51-57.

18. Davidson, D., Wetzel, R., Pennington, D., Ellis, W., and Moore, T., 1985, *In Situ* Treatment Methods for Contaminated Soils and Groundwater. HMCRI Seminar, November 4-5, 1985, Washington, D.C.

19. Dragun, J., 1988: *The Soil Chemistry of Hazardous Materials,* Hazardous Control Research Institute, Silver Springs, Maryland, 458 p.

20. Dunlap, L. E., 1984, Abatement of Hydrogen Vapors in Buildings: In Proceedings of the National Water Well Association and American Petroleum Institute Conference on Petroleum Hydrocarbons and Organic Chemicals in Groundwater—Prevention, Detection and Restoration, p. 504-518.

21. EPA, 1984, Review of Inplace Treatment Techniques for Contaminated Surface Soils, Volume I: Technical Evaluation. EPA-540/2-84-003a.

22. Frankenberger, W. T., Emerson, K. D., and Turner, D. W., 1989, *In Situ* Bioremediation of an Underground Diesel Fuel Spill: A Case History: *Environmental Management,* Vol. 13, No. 3, p. 325-332.

23. Hazaga, D., Fields, S., and Clemens, G. P., 1984, Thermal Treatment of Solvent Contaminated Soils: In Proceeds of the United States Environmental Protection Agency Fifth National Conference on Management of Uncontrolled Hazardous Waste Sites, Washington, D.C., p. 404-406.

24. Hinchee, R. E., Downey, D. C., and Coleman, E. J., 1987, Enhanced Bioreclamation, Soil Venting, and Groundwater Extraction: A Cost Effectiveness and Feasibility Comparison: In Proceedings of the National Water Well Association and American Petroleum Institute Conference on Petroleum Hydrocarbons and Organic Chemicals in Groundwater — Prevention, Detection, and Restoration, November, 1987, p. 147-163.

25. Hinchee, R. E., Downey, D. C., and Beard, T., 1989, Enhancing Biodegradation of Petroleum Hydrocarbon Fuels Through Soil Venting: In Proceedings of the National Water Well Association Conference on Petroleum Hydrocarbons and Organic Chemicals in Groundwater: Prevention, Detection and Restoration, November, 1989, p. 235-248.

26. Hoag, G. E. and Marley, M. C., 1986, Gasoline Residual Saturation in Unsaturated Uniform Aquifer Materials: *Journal of Environmental Engineering,* 112(3), p. 586-604.

27. Jafek, B., 1986, VOC Air Stripping Cuts Costs: Waste Age, Vol. 17, No. 10, October, 1986, p. 66-67.

28. Kerfoot, W. B., 1988, The Soil-Scrub™ Process for Rapid Decontamination of Gasoline-Impregnated Soil: In Proceedings of the Second National Outdoor Action Conference on Aquifer Restoration, Groundwater Monitoring and Geophysical Methods, Vol. I, Las Vegas, Nevada, May, 1988, p. 123-134.

29. Kostecki, P. T. and Calabrese, E. J., 1989, *Petroleum Contaminated Soils,* Vol. 1: Lewis Publishers, Chelsea, Michigan, 357 p.

30. Laney, D. F., 1988, Hydrocarbon Recovery as Remediation of Vadose Zone Soil/Gas Contamination: In Proceedings of the National Water Well Association Second National Outdoor Action Conference on Aquifer Restoration, Groundwater Monitoring and Geophysical Methods, Vol. III, Las Vegas, Nevada, May, 1988, p. 1147-1171.

31. Lee, M. D., Thomas, J. M., Borden, R. C., Bedient, P. B., Ward, C. H., and Wilson, J. T., 1988, Biorestoration of Aquifers Contaminated with Organic Compounds: *Crit. Rev. Environ. Control,* Vol. 18, Issue 1, p. 29-89.

32. Malot, J. J., 1985, Unsaturated Zone Monitoring and Recovery of Underground Contamination: In Proceedings of the National Water Well Association Fifth National Symposium on Aquifer Restoration and Ground Water Monitoring, May, 1985, p. 539-545.

33. Marin, P. R. and Woodhouse, E. G., 1989, Asphalt Batching: An Alternative to Landfilling Disposal of Petroleum Hydrocarbon Contaminated Soil: In Proceedings of the National Water Well Association Conference on Petroleum Hydrocarbons and Organic Chemicals in Groundwater: Prevention, Detection and Restoration, November, 1989, p. 505-518.

34. Marley, M. C. and Hoag, G. E., 1984, Induced Soil Venting for Recovery/Restoration of Gasoline Hydrocarbon in the Vadose Zone: In Proceedings of the National Water Well Association and American Petroleum Institute Conference on Petroleum Hydrocarbons and Organic Chemicals in Groundwater — Prevention, Detection, and Restoration, p. 473-503.

35. Morin, P. R. and Woodhouse, E. G., 1989, Asphalt Batching: An Alternative to Landfill Disposal of Petroleum Hydrocarbon Contaminated Soil: In Proceedings of the National Water Well Association and American Petroleum Institute Conference on Petroleum Hydrocarbons and Organic Chemicals in Groundwater: Prevention, Detection, and Restoration, p. 505-518.

36. Morrow, M. T. and VanDerpool, G., 1988, The Use of a High Efficiency Blower to Remove Volatile Chlorinated Organic Contaminants from the Vadose Zone. A Case Study: In Proceedings of the National Water Well Association Second National Outdoor Action Conference on Aquifer Restoration, Groundwater Monitoring and Geophysical Methods, Vol. III, Las Vegas, Nevada, May, 1988, p. 1111-1135.

37. O'Connor, M. J., Agor, J. G., and King, R. D., 1984, Practical Experience in the Management of Hydrocarbon Vapors in the Subsurface: In Proceedings of the National Water Well Association and American Petroleum Institute Conference on Petroleum Hydrocarbons and Organic Chemicals in Groundwater — Protection, Detection, and Restoration, November, 1984, p. 519-533.

38. Olsen, R. L., Fuller, P. R., Hinzel, E. J. , and Smith, P., 1986, Demonstration of Land Treatment of Hazardous Waste, Superfund 86.

39. Oma, K. H. and Buelt, J. L., 1989, *In Situ* Heating to Detoxify Organic Contaminated Soils: *Hazardous Materials Control*, Vol. 2, No. 2, March/April, 1989, p. 14-19.

40. Overcash, M. R. and Pal, D., 1979, *Design of Land Treatment Systems for Industrial Wastes - Theory and Practice*, Ann Arbor Science Publishers, Ann Arbor, MI.

41. Payne, T. T. and Durgin, P. B., 1988, Hydrocarbon Vapor Concentrations Adjacent to tight Underground Gasoline Storage Tanks: In Proceedings of the National Water Well Association Second National Outdoor Action Conference on Aquifer Restoration, Groundwater Monitoring and Geophysical Methods, Vol. III, Las Vegas, Nevada, May 1988, p. 1173-1197.

42. Roy F. Weston, Inc., 1986, Economic Evaluation of Low Temperature Thermal Stripping of Volatile Organic Compounds from Soil. U.S. Army Toxic and Hazardous Materials Agency Report No. AMXTH-TE-CR86085.

43. Sale, T. and Pitts, M., 1989, Chemically Enhanced *In Situ* Soil Washing: In Proceedings of the National Water Well Association and American Petroleum Institute Conference on Petroleum Hydrocarbons and Organic Chemicals in Groundwater: Prevention, Detection, and Restoration, p. 487-503.

44. Testa, S. M. and Patton, D. L., 1991, Paving Market Shows Promise: Soils, November/December, 1991, p. 9-11.

45. Thomas, J. M., Lee, M. D., Bedient, P. B., Borden, R. C., Canter, L. W., and Ward, C. H., 1987, Leaking Underground Storage Tanks: Remediation with Emphasis on In Situ Biorestoration: EPA/600/2-87/008, p. 144.

46. Thornton, J. S. and Wootan, W. L., Jr., 1982, Venting for the Removal of Hydrocarbon Vapors from Gasoline Contaminated Soil: *Journal of Environmental Scientific Health* A17(1), p. 31-44.

47. Thornton, J. S., 1984, Removal of Gasoline Vapor from Aquifers by Forced Venting: In Proceedings of the 1984 Hazardous Material Spills Conference.

48. Treweek, G. P. and Wogee, J., 1988, Soil Remediation by Air/Stream Stripping: In Proceedings of the Hazardous Materials Control Research Institute Fifth National Conference on Hazardous Wastes and Hazardous Materials, April, 1988, p. 147-153.

49. U.S. EPA, 1988, Cleanup of Releases from Petroleum UST's (Underground Storage Tanks), Selected Technologies: EPA/530/UST-88/001, (PB 88241856).

10 ECONOMIC CONSIDERATIONS OF LNAPL HYDROCARBON RECOVERY

"If you fail to plan, you plan to fail"

10.1 INTRODUCTION

Recovery of NAPL from an aquifer restoration site is an essential part of the clean-up effort. Removal of the floating LNAPL layer is almost always required prior to the initiation of other restoration- and remediation-related activities (i.e., enhanced bioactivity, vapor extraction, etc.) and responding to other environmental issues (i.e., hydrocarbon-affected soils, vapors, or dissolved hydrocarbon in groundwater).

In principal, the recovery of LNAPL is similar in mechanical operation to production of a low-pressure, water-drive reservoir. Almost all documented petroleum remediations have been characterized by subsurface conditions under water table conditions (i.e., the top surface of the fluids are at atmospheric pressure). A few cases of confined aquifer situations have been reported (Trimmel, 1987) and although the mechanical recovery procedures are slightly different, the economic considerations are similar.

Detailed planning is necessary to ensure that the field work meets predetermined goals with regard to administrative requirements, the development of recovery facilities, and optimized production. At any specific site, the goals may include any or a combination of the following:

1. *Meeting regulatory compliance.* In many situations, recovery efforts are driven by statutory forces, which are result oriented and are not particularly oriented toward any specific site. Most projects of this variety have specified quality time constraints and the economic considerations are secondary.
2. *Production of greatest economic return cash flow.* Many older petroleum refining operations have experienced individual spills or continual small-scale leakage over a prolonged period of time. A large percentage of these leaks occur from underground sources, and because of geological setting, have not been apparent. A loss of .5% may not be detected by the operator, but can accumulate significantly over an extended period of time. For example, a refinery with production of 60,000 barrels/d with undetected losses of 300 barrels/d would have accumulated leakage of a total of over three million barrels during a 30-year period. Under favorable geological and hydrogeological conditions, this product may be an economic asset. Careful management of recovery operations can often produce a positive cash flow (or at least significantly reduce costs), as well as provide positive environmental benefits.
3. *Maintenance of current commercial status quo.* At a few locations, the decision has been made to continue operation of the facility without disruption, so long as LNAPL product does not exit the boundary of the facility. Hydraulic containment of the aquifer is the procedure that is usually selected for these sites. A system including recovery wells and injection wells can often be operated to balance the subsurface flux so that product loss equals product recovery at a minimal cost.
4. *Accomplishment of "best environmental restoration".* A thorough understanding of the geological and hydrogeological setting, including the LNAPL recovery potential, can lead to development of a total remediation plan. Under this type of goal, overall clean-up of the contaminant is the primary goal. The use of variable hydraulic flow and flushing techniques, combined with the utilization of both natural and enhanced biodegradation, will ultimately result in a "best case" total restoration. Investigative and management effort for this type of project is extensive.

The following discussion is directed toward evaluating the costs of operating LNAPL recovery systems. The principles presented are equally applicable to all types of LNAPL recovery situations.

10.2   PRELIMINARY CONSIDERATIONS

Each recovery site has its own individual site specific characteristics. Subsurface stratigraphy and other geologic considerations, depth to water table, permeability, aquifer(s) thickness, size of product pool, physical characteristics of the product, effects of product weathering, gradient, source of leakage, and groundwater quality are all factors of importance in the design and operation of a LNAPL recovery system.

Initial investigations must be adequate to provide sufficient information to support the design and operation of the recovery system. Progressive and flexible testing procedures during the investigative phases will require sufficient hydraulic testing to characterize the product producing areas and to evaluate the approximate quantity and rate of recovery. The data procured should also provide guidance in assessing how many recovery and monitoring wells are required and where to best situate them.

Determination of initial recovery well (or trench) locations is an important design parameter. Floating LNAPL product tends to move in the direction of overall groundwater flow, as determined by the water table gradient. As a well or trench is pumped the fluids (water and/or oil) migrate toward the area of lower pressure to fill the void. A cone of depression develops that extends outward. The fluid surface exhibits a rapid slope near the well, diminishing to a very low gradient at a distance.

Floating product migrates toward the well at a rate proportionate to the hydraulic gradient. As production wells deplete oil reserves in the immediate vicinity, the oil must travel farther, over lesser gradients, to reach the well. The end result is reduced oil production. Construction of new wells at more advantageous intermediate locations will revitalize production. An alternative procedure is to install an injection well between recovery wells. The increase in the hydraulic gradient caused by the injection well would then stimulate oil flow toward the recovery wells.

During the design phase, all of the data derived from the hydraulic characterization are evaluated for use in the selection of recovery pumping equipment and for the determination of the most appropriate subsurface fixtures (whether wells, trenches, or drains, etc.). A variety of generic recovery scenarios may be appropriate to optimize product recovery. If the product thickness is sufficient, the viscosity low, and the formation permeable, a simple pure-product skimming unit may be the best choice. Other combinations of permeability, geology, and product quality will require more active systems, such as one-pump total fluid, or two-pump recovery wells.

The selection of the recovery equipment should be based on functionability, capital cost, operational cost, and ease of movement between individual locations. A well-managed recovery system will include routine relocation of equipment in response to recovery needs. Flexibility of usage is important because the capital cost of the recovery equipment represents a significant percentage of the project cost.

10.3 ECONOMIC FACTORS

All projects involving any significant quantity of LNAPL product recovery require the consideration of economic factors. Careful planning to optimize each project phase can lead to the lowest cost of operation and can occasionally generate positive cash flow, while currently accomplishing aquifer restoration. A basic premise in this discussion is that the recovered LNAPL is suitable for reuse (i.e., as refinery feed stock or fuel for incinerators). Products unsuitable for resale only add to the debit side of the economic equation.

Debits associated with these projects are the costs of

- Investigation and testing
- Evaluation and design
- Construction of field facilities
- Operation and maintenance
- Money invested (time value)

Credits resulting from the recovery include returns of

- Value of product recovered
- Time value of money from sold products

A summation of the credits and debits describes the economic status of the project. A carefully planned and executed project results in the greatest possible credit balance for the longest period of time.

The following example demonstrates the principles involved in the optimization of a project. A fairly large older refinery situated on an alluvial flood plain had been in continuous operation for 35 years. Leakage and spillage were common, in almost unnoticed quantities (<0.5% by volume), resulting in a significant accumulation of product on the shallow water table over the course of operation. The intended goal of the recovery program was to accomplish the "best environmental restoration" at the least cost. For convenience, all costs and returns are presented in terms of value per barrel. A graphical presentation of the cumulative costs (in barrels) vs. time and a summation of the volume of recovered product vs. time is shown in Figure 10.1. The following costs were assumed:

- Investigation and testing phase cost 3,000 barrels over a 2-month period
- Evaluation and design cost 1,000 barrels over a $1^1/_2$-month period
- Construction of field equipment cost 8,000 barrels over a $1^1/_2$-month period
- Operational costs remained constant at 1,500 barrels per month from initiation to completion of the recovery phase

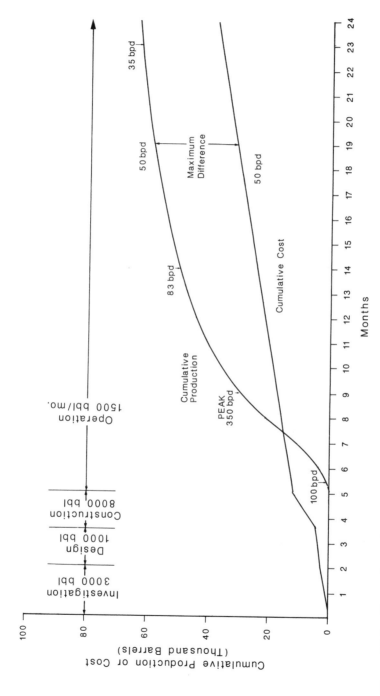

Figure 10.1. Example project of LNAPL recovery, cumulative costs, and production vs. time.

Table 10.1   Recovery of LNAPL With Time

Month	Production (Barrels per Day)
5	100
9	350 (peak)
14	83
18	58
23	35
45	7 (projected value)

Recovery started in the fifth month and continued until completion. Recovery rates associated with Figure 10.1 are summarized in Table 10.1. The example illustrated in Figure 10.1 follows the typical case in which the recovery rate increases to a peak rate, then rapidly declines toward an asymptotic curve over several months. The use of cumulative curves allows a comparison of the relative positions of both cost and production, as illustrated in Figure 10.2. Prior to $7^1/_2$ months, the project operated at a cumulative loss. After that time, the cumulative recovery was greater than the accumulated cost. This positive situation continued until month 45 (projected off the edge of the chart), where the production costs equaled the value of the product recovered. The distance between the two curves was at a maximum near month 18, when the product was being recovered at a rate of 50 barrels/d at a cost of 50 barrels/d. Beyond this time, the return rate is negative. A project planned to focus on positive cash flow would end at this point, although an additional 25 months would pass before the total value of recovered product equaled costs. Since this project was planned to accomplish long-term remediation, funds earned during the positive cash-flow period were retained to finance subsequent longer term phases of the project. This planning resulted in a greatly reduced overall cost.

Optimization involves careful planning during all phases. The investigation phase is "adequate", with sufficient drilling and testing to characterize the site and to identify optimum locations for recovery wells. Interviews with refinery maintenance staff can prove invaluable in delineating potential source areas. Existing boring logs also provide data on subsurface geologic and hydrogeologic conditions and permeability.

During the design phase, the primary goal is to specify reliable equipment that is easily maintained and can be operated with minimal energy. Downtime, while awaiting a repair specialist or the arrival of unique parts, can cost more than a slightly less efficient system that is easily repaired by on-site staff. Installation and activation of the field equipment should be conducted in a manner that initiates the recovery of LNAPL as soon as practical. As each additional well is installed, piping, energy for pumping, control units, tankage, and other necessary

Time-Rate of Production

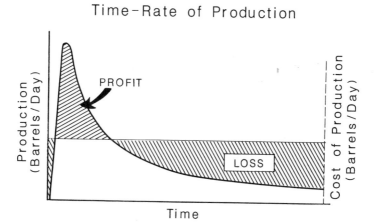

Figure 10.2.   Time-rate of production vs. operating cost.

utilities should also be ready for use. Proper prior planning and careful coordination are critical to efficient operations.

Operation of a system on a continuing basis requires regular monitoring and maintenance. This phase represents the largest cost items of a long-term project; however, routine adjustments of pumping units, coupled with preventative maintenance, will result in lower long-term costs.

10.4   PROJECT PLANNING AND MANAGEMENT

The preceding paragraphs describe a scenario related to a single-phased project. The initial well system was operated as a single unit throughout the duration of recovery. Although this technique was appropriate for that site, alternative programs may often be better suited for other locations. At sites that have extensive "reserves", it has often been demonstrated that a phased approach can improve recovery efficiency. This procedure involves optimizing the use of recovery equipment by progressive replacement of high-capacity oil pumping equipment with equipment of lesser capacity as production declines in each well. A typical recovery curve for a well, or system of wells, is shown in Figure 10.3. Because prime areas were selected and exploited first, the peak rate of recovery for each phase tends to be less for each succeeding phase. Also, the break-even line increases as operational costs for each new phase is added. After several phases have been added, the cost of operation will eventually meet or exceed recovery rates.

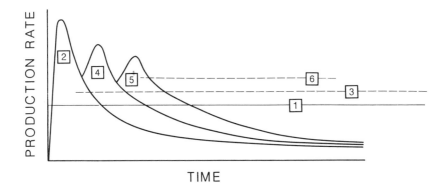

TIME

1 1st Phase Break-Even Line Includes
Costs Involved With Site Investigation
and Equipment and Energy and Labor and
Maintenance, etc.

2 Production From Phase 1.

3 Break-Even Line for Phase 2 and Phase
1. Includes Incremental Cost of Installation of Additional Wells.

4 Production From Phase 3 and Phase 1.

5 Production From Phases 3 + 2 + 1.

6 Break-Even Line for Phases 1 + 2 + 3.

Figure 10.3.   Time-production rate vs. operational cost for a multiphased project.

10.5   ESTIMATING RESERVES

Preliminary estimates of LNAPL made during the investigative phases of a project are usually based on the results of short-term pumping tests, approximations of actual LNAPL thickness based on gauging data generated from monitoring wells, and other approximate data. The numerical estimates based on this short-term information are often adequate enough to define the physical parameters sufficiently for preliminary design purposes. However, long-term reserve estimates and the economics of production can only be completed after at least part of the recovery system has been operating long enough to establish a distinctive production pattern.

A simplified procedure for the analysis of production decline curves was developed by Gentry (1972). This method was initially intended for the evaluation of individual oil well production; however, it also provides reasonable estimates when applied to multiple well LNAPL recovery systems. This analytical method is applicable to most types of decline curves, whether they tend to follow exponential, hyperbolic, or harmonic forms. The following general differential equation is applicable to all forms of decline curves.

$$D = Kq^n = -\left(\frac{dq}{dt}\right)q, \qquad (10.1)$$

where D = decline; K = operative constant; dq = differential rate of production; dt = differential time; q = production rate; and n = decline exponent.

Specific equations that are all solutions for the general equation are presented in Table 10.2. Because each of these equations contains two unknowns, solution requires the input of data from external sources. Necessary values derived from the relationships of field-measured data are applied to the theoretical equation to derive a solution. Any set of consistent measurement units may be used. Field measured data are

q_i = initial production rate (units per time).

q_t = measured production rate at time t after initial peak production.

Q_t = accumulated production rate between q_i and time t.

Use of these equations to predict future production from a recovery project is described by the following example. An abandoned refinery property is being dismantled and the underlying aquifer remediated. Substantial LNAPL product accumulations occurred overlying the fine silty sand aquifer. Preliminary investigation indicated that a four-well system would effectively remove most of the product within a reasonable time at a modest cost. The production rate over time is illustrated in Figure 10.4. Peak production occurred on the 78th day of operation, then declined. Final measurement occurred on day 141.

Step 1. Determine which decline equation most closely defines the actual curve.

Input data: $q_i = 21$ bpd (barrels/day)

$q_t = 14$ bpd

$t = 63$ days

$Q_t = 1015$ barrels (actual measured)

Exponential equation:

$$D_i = \frac{In\left(\dfrac{q_i}{q_t}\right)}{t} \qquad\qquad Q_t = \frac{(q_1 - q_t)}{D_i}$$

$$D_i = .00644 \qquad\qquad Q_t = 1094 \text{ barrels}$$

Table 10.2   Summary of Mathematical Solutions for Production Decline Curves

Decline Exponent	Type of Decline	Rate/Time Relationship	Rate: Cumulative Relationship	$D_i t$ Relationship	$Q_t/q_t t$ Relationship
$n = 0$	Exponential	$q_t = q_i e^{D_i t}$	$Q_t = (q_i - q_t)/D_i$	$D_i t = \ln(q_i/q_t)$	$Q_t/q_i t = \dfrac{1-(q_i/q_t)^{-1}}{\ln(q_i/q_t)}$
$D < n < 1$	Hyperbolic	$q_t = q_i\left(1 + nD_i t\right)^{-1/n}$	$Q_t = \dfrac{q_i^n}{(1-n)D_i}$ $\cdot\,(q_i^{1-n} - q_t^{1-n})$	$D_i t = \dfrac{(q_i/q_t)^n - 1}{n}$	$Q_t/q_i t = \dfrac{1-(q_i/q_t)^{n-1}}{(q_i/q_t)^n - 1}$ $\cdot\left(\dfrac{n}{1-n}\right)$
$n = 1$	Harmonic	$q_t = q_i(1+D_i t)^{-1}$	$Q_t = (q_i/D_i)\ln(q_i/q_t)$	$D_i t = (q_i/q_t) - 1$	$Q_t/q_i t = \dfrac{\ln(ql/q_t)}{(q_i/q_t) - 1}$

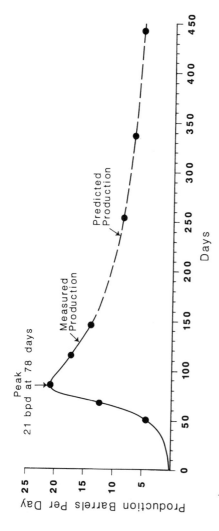

Figure 10.4. Example of project time-reduction rate.

Hyperbolic equation:

Assume n = 0.5 for trial

$$D_i = \frac{\left(\dfrac{q_i}{q_t}\right)^n - 1}{nt}$$

$$Q_t = \frac{q_i^n}{(1-n)D_i}\left(q_i^{1-n} - q_t^{1-n}\right)$$

$$D_i = .00713$$

$$Q_t = 1077$$

Harmonic equation:

$$D_i = \frac{\left(\dfrac{q_i}{q_t}\right) - 1}{t}$$

$$Q_t = \frac{q_i}{D_i}\ln\left(\frac{q_i}{q_t}\right)$$

$$D_i = .00794$$

$$Q_t = 1070$$

The harmonic equation provides the closest answer (within approximately 5%), therefore, that equation is most representative. By use of the rate/time relationship for the harmonic equation, the recovery curve may be projected to predict recovery rates at future dates after peak production.

$$q_t = q_i\left(1 + D_i t\right)^{-1}, \tag{10.2}$$

where at 173 d $q_t = 8.87$ bpd; at 258 days $q_t = 6.91$ bpd; and at 365 days $q_t = 5.40$ bpd. As this is an asymptotic curve, it will approach zero at infinity. In practice, the practical production limit is reached when the actual LNAPL layer is thin enough that it does not migrate freely. Each combination product type, soil variety, and wetted condition (oil-water saturation) behaves differently and must be considered separately. An estimate of total production from this system at 365 d after peak can be determined by use of the rate/cumulative relationship:

$$Q_t = \frac{q_i}{D_i}\ln\left(\frac{q_i}{q_t}\right)$$

$$Q_{1365} = 3590 \text{ barrels} \tag{10.3}$$

This example demonstrates that it is possible to make reasonably accurate predictions of production after the pattern of production is established. However, caution must be exercised by the professional when developing these data. The example cited occurred during a period of time when the water table at the facility was within "normal" ranges. If the water table had risen, or fallen substantially (for any reason), the pattern of production would have changed, and the calculations would not be considered reliable. Estimation of reserves determined by these methods can be fairly reliable if the data used are based on adequate and regular measurements, and are applied with a reasonable measure of professional judgment.

As in any scientific or engineering endeavor, the quantity and validity of input data determine the accuracy of prediction. Frequent gauging of fluid levels in monitoring wells, flow rates, and oil-water ratios, in conjunction with proper quality control, can lead to accurate estimates that support proper project performance.

REFERENCES

1. Arps, J. J., 1956, Estimation of Primary Oil Reserves: *American Institute of Mining and Metallurgical Engineers Transactions,* Vol. 207, p. 182-191.
2. Gentry, Robert W., 1972, Decline-Curve Analysis: *J. Petrol. Technol.*, p. 38-41.
3. Slider, H. C., 1968, A Simplified Method of Hyperbolic Decline Curve Analysis: Journal of Petroleum Technology, p. 235-236.
4. Trimmel, M. L., 1987, Installation of Hydrocarbon Detection Wells and Volumetric Calculations within Confined Aquifers: In Proceedings of the National Water Well Association and American Petroleum Institute Conference on Petroleum Hydrocarbon and Organic Chemicals in Ground Water - Prevention, Detection and Restoration, Vol. I, November, 1987, p. 225-269.
5. Trimmel, M., Winegardner, D., and Testa, S. M., 1989, Cost Optimization of Free Phase Liquid Hydrocarbon Recovery Systems: In Proceedings of the Hazardous Materials Control Research Institute Conference on Hazardous Waste and Hazardous Materials, April, 1989.

11   CASE HISTORIES

"Ignorance is no defense"

11.1   INTRODUCTION

The selection of an approach or the combination of approaches to NAPL recovery and ultimately aquifer rehabilitation and restoration is dependent on numerous factors, as discussed in the previous chapters. The case histories presented below reflect different remediation approaches in response to varying geologic and hydrogeologic conditions, and site-specific requirements. The systems to be discussed include:

- Vacuum-enhanced suction-lift well-point system
- Rope skimming system with bioremediation
- Vacuum-enhanced eductor system
- Combined one- and two-pump system with reinjection (and vacuum-enhancement)

11.2   VACUUM-ENHANCED SUCTION-LIFT WELL-POINT SYSTEM

Near-shore facilities are characterized by shallow groundwater conditions. The occurrence of LNAPL product on the water table presents the need for immediate containment and continued recovery of the product to abate degradation of groundwater quality, hydrocarbon vapor migration, and discharge of product to surface waters. Thus, the need for immediate as well as long-term containment and recovery is required.

The pumping system utilized for such a facility was a pneumatically operated, double-diaphragm, suction-lift pump to withdraw groundwater and mobile LNAPL from the shallow water table aquifers. Applicable over a wide range of hydrogeologic environments with depths to water ranging up to 22 ft, the system can also be used to pump from aquifers of varying characteristics, ranging from low-permeability clay and silt to higher yielding sand formations. The system has been found to be generally compatible with existing site uses, such as refineries, terminals, and gasoline stations. These types of sites also offer favorable characteristics, such as pneumatic operation availability and intrinsic safety. Practical aspects include off-the-shelf availability of proven equipment and the ability to induce additional recovery through the application of a vacuum to the recovery wells.

One of several sites that were found to be suitable for a double-diaphragm suction-lift well-point pumping system was a bulk liquid marine terminal located in Los Angeles Harbor. This facility had been a marine terminal since the early 1920s and is still used for the transfer of petroleum products to and from ships and onshore facilities. Onshore facilities include multiple subsurface petroleum product pipelines to inland facilities and an above-ground storage tank farm with several existing tanks of varying sizes. The initial remedial response was initiated when LNAPL product seepage was noted discharging into harbor waters.

Subsurface conditions were explored by the hand augering of 26 exploratory borings and 61 additional borings to be used for monitoring or recovery well installation. Due to the close proximity of the site to surface water and the shallow depth to groundwater, the maximum depth drilled was approximately 15 ft. The surficial materials encountered across the site were comprised primarily of hydraulically emplaced uncontrolled fill consisting of fine sand and silty sand. The fill materials appeared to be reworked local materials, possibly dredgings from the adjacent harbor, since shell fragments were observed in many of the borings. The base of the fill was inferred to be coincident with the top of a thin, laterally discontinuous clay layer generally encountered between 2 and 5 ft below the existing ground surface. It appeared that the natural land surface prior to emplacement of fill was fairly irregular, such that fill thicknesses varied from 2 to 10 ft across the site, being generally greater along the shoreline behind the rip-rap bulkhead. Below the fill were natural deposits of marine sand, silt, and clay that are typical of an estuarine depositional environment.

Groundwater was encountered under water table conditions at depths ranging from 6 to 8 ft below the local ground surface. Groundwater flow was perpendicular to the shoreline and was generally towards the open water of the harbor, with a gradient of approximately 0.02 feet per foot. Shallow groundwater occurred both within the fill and underlying natural sediments. Grain-size distribution analyses performed on soil samples retrieved from the water-bearing strata indicated that the aquifer materials were predominantly fine to medium sand.

Pumping test data yielded an estimated aquifer transmissivity of 100 gallons per day per foot (gpd/ft) and a resultant permeability of approximately 2 ft/d (7.4 × 10 cm/s). This was consistent within the general range of permeability expected for the soil types observed.

The approach to containment of LNAPL discharge to the harbor followed the concept of a well-point dewatering system for interception of groundwater discharge. The discharge of groundwater and LNAPL hydrocarbon product to the adjacent harbor is illustrated in cross-section in Figure 11.1. Interception of product discharge to the harbor was accomplished by the pumping of groundwater, resulting in a reversal of flow direction. This reversal reflects the lowering of the water table below that of the adjacent harbor water level.

Apparent LNAPL thicknesses measured in monitoring wells ranged up to 7 ft. During the subsurface characterization phase, three distinct petroleum product types were evident, which had apparently coalesced into one LNAPL pool. Product samples retrieved indicated characteristics of gasoline, kerosene, and crude oil. Despite the variation in product types, pumping system design and construction was similar throughout the site, with no significant effect on system efficiency and effectiveness.

The containment of groundwater flow was initially accomplished through the installation and pumping of 10 recovery wells, of which 5 were oriented parallel and adjacent to the shoreline. Based on pumping test results, recovery wells were installed generally on 60-ft centers. In subsequent phases, additional recovery wells were installed further inland within the tank farm area to intercept and recover product prior to migrating to the shoreline area.

Data generated through a year of operation of the recovery system was analyzed by a variety of means to provide both qualitative and quantitative indications of the overall system effectiveness. Seepage of LNAPL from the shoreline to the harbor waters had manifested itself as staining of the rip-rap and a visible sheen of product on the water surface. No specific quantification was made of the rate of seepage prior to the initiation of hydrocarbon recovery. Regardless, after approximately 3 months of operation, no visible seepage was evident. In addition, visible product evident on the rip-rap within the zone of tidal fluctuation had been eliminated. Estimates of the volume of product recovered were developed from the pumping rate and the percent product data. The average pumping rate ranged from 0.68 to 1.87 gallons per minute (gpm). With an average of three wells per pump, the estimated pumping rates for individual wells ranged from 0.22 to 0.62 gpm. The average percent product in total fluids pumped for individual pumps ranged from 7.6% to 22%, with no observable difference in the recovery rates for the different product types present. Overall, an estimated 569,000 gal (13,548 barrels) of LNAPL product was recovered during the initial year of recovery from 33 recovery wells manifolded to 11 pumps.

Reductions in the apparent LNAPL thickness in the subsurface have been

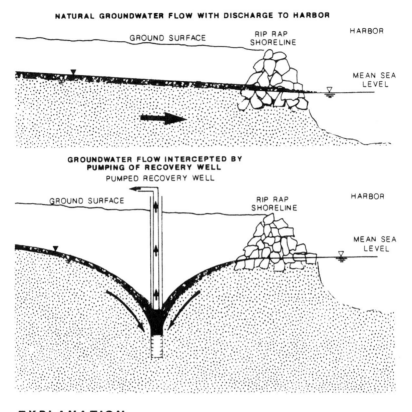

EXPLANATION

▽ *Groundwater level*

▼ *Mobile LNAPL level*

▩ *Mobile LNAPL*

◄─ *General Direction of Groundwater Flow*

Figure 11.1.   Cross-sectional view of LNAPL containment system under shallow groundwater conditions at a harbor facility.

effected over most of the site. In general, the apparent thickness after 1 year was reduced 30–40% to those apparent thicknesses measured prior to recovery. The reduction in apparent thickness in a representative monitoring well with time is shown in Figure 11.2. However, in one area of the site, hydrocarbon thicknesses actually increased as a result of pumping, reflecting redistribution of the product in response to the creation of a cone of depression in an area of relatively higher permeability. The installation of additional recovery wells in this localized area

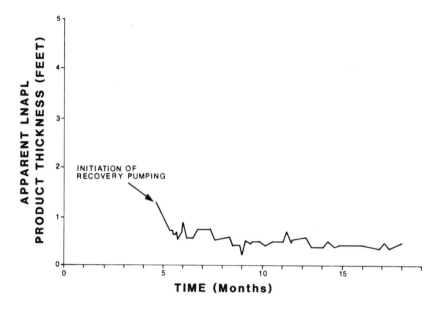

Figure 11.2. **Apparent LNAPL thickness decrease with time due to the initiation of recovery operations.**

effected a rapid decrease in the apparent LNAPL thickness. This increase and the subsequent decrease in one monitoring well is graphically illustrated in Figure 11.3.

The effectiveness of a suction-lift well-point pumping system for LNAPL recovery under shallow water table or perched conditions is evident from experience gained in this case history. The primary effectiveness is attributed to the inherent capabilities of suction-lift equipment, combined with the practical aspects of using such equipment at petroleum-handling facilities. Due to the flammable and/or explosive nature of hydrocarbon products, pumps that are driven by compressed air are inherently safer than pumps requiring electrical power. In addition, many petroleum-handling facilities, including refineries, terminals, and gasoline stations have existing compressors that can be used to supply the necessary air volume for pump operation. The pumping equipment and controls (valves) are readily available through many industrial supply distributors. Since double-diaphragm pumps have multiple uses, there are many brands and sizes to select from to suit anticipated pumping requirements.

The use of less sophisticated electronic controls lends to significantly reduced equipment costs. Double-diaphragm pumps are self-priming at a lift of 22 ft. If suction is not broken, these pumps may recover fluids from as deep as 27 ft. In addition, the pumps will operate against pressures of up to 110 lbs/in.[2] Double-diaphragm pumps can also handle a wide variety of petroleum products, ranging

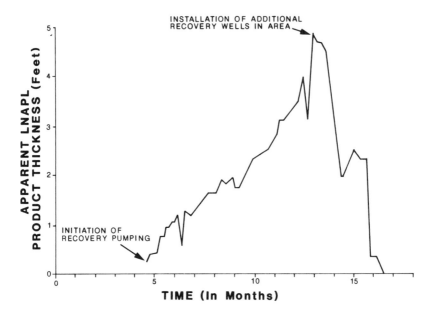

Figure 11.3. Apparent LNAPL thickness fluctuations with time reflecting installation of additional recovery wells.

from gasoline through heavy crude oil, and are not significantly affected by the pumping of solids such as sand or silt (<1/8-in. diameter).

Double-diaphragm suction-lift pumps also have advantages relative to varying subsurface hydrogeologic characteristics encountered at different sites. These include:

- Wide range in pumping rates
- Can be pumped dry
- Vacuum-assisted recovery

The pumps can sustain varying pumping rates, ranging from virtually no discharge to tens of gallons per minute. Pumping rate is easily controlled by valving the air supply intake or discharge from the pump. In low-permeability formations, where the pumping rate exceeds the well(s) yield, double-diaphragm pumps are not damaged by pumping without available fluid. A vacuum will be maintained until the fluid level in the well(s) recovers, and recovery of fluids resumes. By sealing the well head to air entry, the vacuum that develops within the well enhances fluid flow to the well by simulating additional drawdown, thereby increasing the effective recovery of the well.

Overall, the pumping system has been effective in eliminating surface water discharge of LNAPL product, reducing apparent product thickness to 30–40%

of preexisting conditions, while recovering a significant volume of product for recycling.

11.3   ROPE SKIMMING SYSTEM

An abandoned refinery in the Midwest is located near a suburb of a medium-sized city. The site encompasses about 60 acres and has remained unoccupied since surface features at the refinery were removed in the 1960s. During the late 1970s adjacent residents began to complain that their shallow irrigation wells produced contaminated water that smelled like gasoline.

Subsurface geologic and hydrogeologic conditions were evaluated based on excavations of test pits (on 100-ft centers), installation of monitoring wells, and by conducting *in situ* rising-head slug tests and pumping tests. The site is located within a glaciated region and is immediately underlain by an unconsolidated, stratified, sequence of silt and sand to a depth of approximately 50 ft below ground surface. These relatively coarse-grain deposits overlie deposits of highly plastic inorganic clay, which acts as a lower confining unit. Geologic contacts between the upper sand and silt, and lower clay deposits are gradational. A site facility layout map showing well locations is shown in Figure 11.4. A hydrogeologic cross-section is shown in Figure 11.5.

Groundwater occurs under shallow water table conditions at an approximate depth of 5–8 ft below ground surface with a southwest gradient of .004 feet per foot. The configuration of the water table showing the general direction of groundwater flow is shown in Figure 11.4. Hydraulic conductivity based on *in situ* rising-head slug tests and pumping tests was on the order of 4.3×10^{-3} cm/s (92 gpd/ft^2). Groundwater supplies for industrial purposes are typically derived from the upper 75 ft of the glacial deposits.

Chemical analysis of water samples collected from monitoring wells and adjacent ditches was performed for benzene, toluene, ethylbenzene, xylene, total organic carbon, chromium, and lead. The results of the analysis presented a complicated chemical distribution in a dynamic environment. The concentration of each of the analyzed compounds ranged from less than 2 µg/l to nearly 1 mg/l with a few higher levels reported. The area of primary concern was one well with benzene reported at 1400 µg/l. The most valuable indicator of ground water contamination was total organic carbon content (TOC). This parameter ranged from 7 mg/l in the upgradient wells to 40 or 60 mg/l in the center of the site. TOC distribution across the site is presented in Figure 11.6.

LNAPL product was found to be distributed over the central portion of the site with a measured apparent thickness of approximately 0.2–0.5 ft. Concluding that additional work was warranted at the site, two goal-oriented objectives were developed:

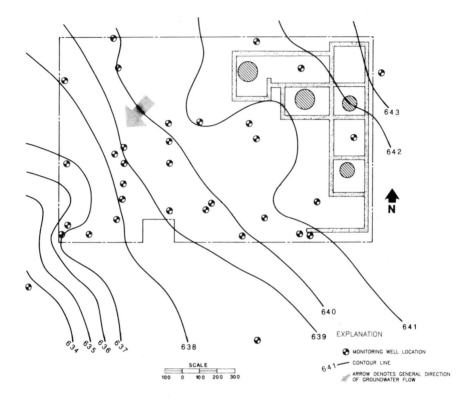

Figure 11.4.   Refinery site facility layout map showing well locations and the configuration of the water table.

1. Consideration of on-site clean-up to prepare the site itself for other industrial uses
2. Determination of the extent of off-site groundwater contamination

Early in the investigation phase it became apparent that removal of the LNAPL was an essential part of the aquifer restoration program. A pilot recovery well was constructed to determine the feasibility of recovery of LNAPL. This 48-in. diameter, backhoe-excavated well was equipped with a submersible water pump and a LNAPL recovery skimmer. The water pump created a cone of depression, while the skimmer recovered mobile LNAPL that migrated toward the well. Water produced was discharged onto the ground surface 200 ft from the well. Relatively high soil permeability allowed the water to recharge quickly so that no significant ponding occurred.

This pilot recovery produced significant product initially, but decreased rapidly due to a limited area of influence. Based on the results of this test, it was concluded that the use of individual wells to recover the mobile LNAPL was not

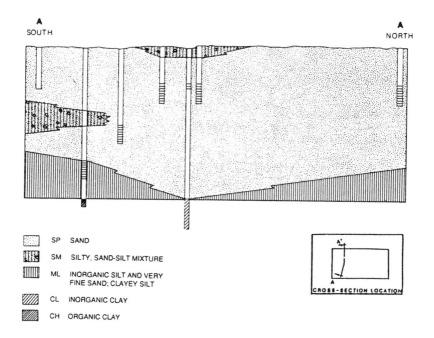

Figure 11.5. Hydrogeologic cross-section of refinery site facility.

an acceptable option. In addition, the number of wells, pumps, and the effort of maintenance were considered to be restrictive.

An alternative to installing individual wells with limited areas of influence was to construct a system of open drainage ditches with a dendritic pattern spread over an area of LNAPL occurrence. A single submersible pump, at the end of the trench, was provided to assure flow toward the recovery area. A single "rope skimmer" recovered the LNAPL as it migrated toward the pump. A schematic sketch of the trench recovery system is shown in Figure 11.7. The general layout of the trenches is shown in Figure 11.8. Water from the lower end of the trench was pumped to an upgradient recharge pond. The increased hydraulic gradient with enhanced mobile hydrocarbon recovery rates prevented the system from being dewatered.

Trenches were constructed using conventional earth-moving equipment. During construction, air monitoring was continued to ensure that explosive conditions were not present. However, odors persisted from the freshly excavated soils but disappeared within a few days. Operation of this trench system continues throughout frost-free seasons and has proved very successful. Initially, approximately 5 barrels per day (bpd), which has since diminished, was recovered. When the product is essentially gone, the trenches will be refilled with the original excavated material.

Clean-up of residual LNAPL in the shallow soils was considered to be a

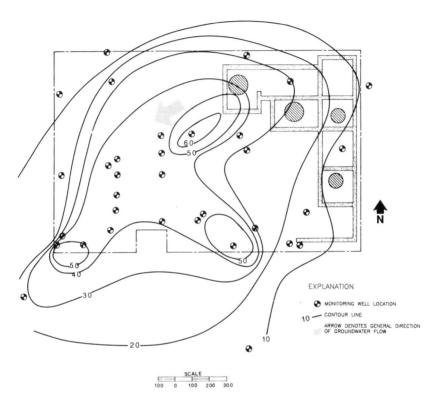

Figure 11.6.   Total organic carbon distribution in shallow groundwater beneath refinery site.

necessary part of the overall remediation program. Enhanced biological activity was initially considered. A laboratory demonstration proved that naturally occurring local oil bacteria could be used to accomplish this task. A pilot laboratory study was conducted to investigate the scale-up treatment parameters for operation in the field.

A 10-yd^3 soil sample was excavated from the site, blended, and characterized for initial hydrocarbon content and nutrient content. The reactor was filled with soil compacted to field density (Figure 11.9). The tank at the bottom was filled with water nutrients and surfactants. Water from this tank was sprayed over the top of the soil at a rate that maintained aerobic conditions.

Initially, a significant amount of product was released from the soil that required additional air to be pumped into the well points to maintain favorable growth conditions. After 105 days of operation, more than 87% of the total aliphatics and 89% of the total aromatics were removed.

Based on the successful reactor study, a 1-acre field demonstration was

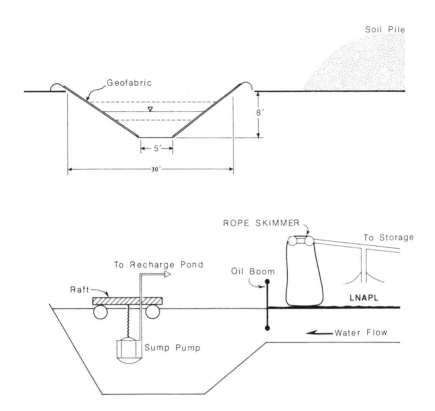

Figure 11.7.   Schematic of trench recovery system beneath refinery site.

performed to evaluate the potential for full site remediation of the unsaturated zone. The selected area was surveyed to be exactly 1 acre (circular), carefully sampled to determine the quantity of residual hydrocarbon, and place counts were made to identify the species of soil microbes present. A single recovery well was installed and tested in the center of the plot. The purpose of this well was to create a drawdown cone extending to the boundary of the plot to ensure collection of all fluids infiltrating during the test.

A 50,000 gal tank was installed adjacent to the test area. This tank was equipped with an aerator, nutrient feed equipment, and a submersible discharge pump. The physical layout of this system was very similar to that shown in Figure 11.9. Make-up water from the recovery well was pumped into the tank at a rate of approximately 38 gpm. Soluble nitrogen and phosphorous fertilizer compounds were metered into the tank and mixed by the action of the aerator. The submersible pump was used to spray irrigate the treatment plot with the enriched water. The average application was approximately 2 in./d over the test area (slightly less than the vertical infiltration rate of the shallow soil). During

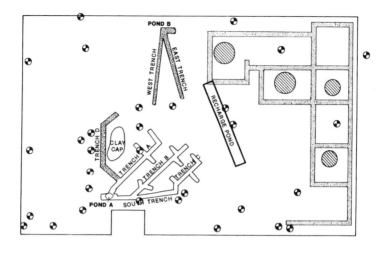

Figure 11.8. **General layout of monitoring wells and trenches at refinery site.**

rainstorms, it was necessary to discharge some water to an off-site holding pond; however, this did not appear to significantly affect the long-term test results. After one summer of operation, the field demonstration confirmed that the residual hydrocarbon could be reduced by this procedure. In the following years, plans were made to expand the treatment to larger areas.

When off-site contamination was confirmed, computer modeling was used to estimate and predict future dispersion and transport. Total organic carbon was the indicator parameter selected. The use of a numerical model allowed projections to be made without causing public concern during field work. The model selected was the plasm model sequence. Hydraulic parameters were initially based directly on the data derived from field work. This original data was manipulated somewhat in order to calibrate the model to match the observed field conditions. The initial distribution of TOC is shown in Figure 11.10. Further runs were made to project the contaminated spread during the following years. A map of predicted TOC distribution after 6 years from the initial study is shown in Figure 11.11. Follow-up sampling indicated that the projections were realistic.

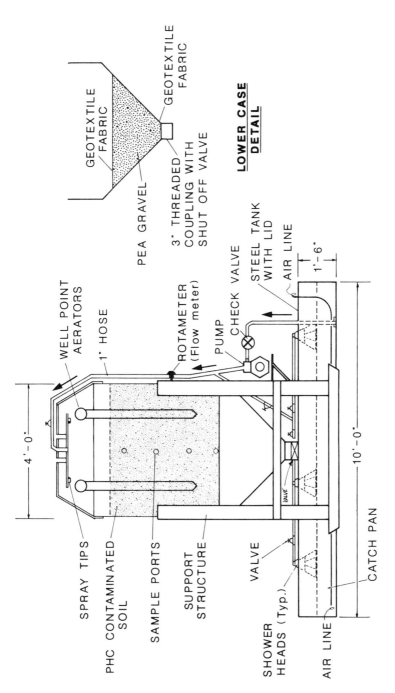

GEOTEXTILE FABRIC

GEOTEXTILE FABRIC

PEA GRAVEL

3" THREADED COUPLING WITH SHUT OFF VALVE

LOWER CASE DETAIL

WELL POINT AERATORS

1" HOSE

ROTAMETER (Flow meter)

PUMP

CHECK VALVE

STEEL TANK WITH LID

AIR LINE

1'-6"

10'-0"

4'-0"

SPRAY TIPS

PHC CONTAMINATED SOIL

SAMPLE PORTS

SUPPORT STRUCTURE

VALVE

SHOWER HEADS (Typ.)

AIR LINE

CATCH PAN

VALVE

Figure 11.9.   Schematic of soil-nutrient reactor used for biotreatment pilot study.

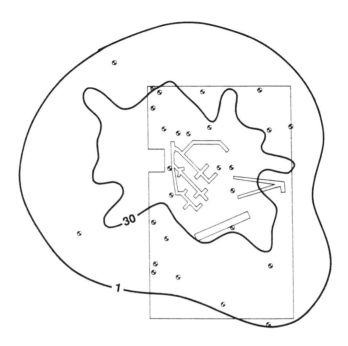

Figure 11.10.   Initial total organic carbon distribution in shallow groundwater beneath refinery site.

Remediation currently continues at this site. Recovery of LNAPL will continue for the next few years. Further expansion of the biological treatment system
is planned to follow the LNAPL recovery effort. Off-site groundwater remediation has been demonstrated to be unfeasible due to the volume involved and the
limited rate of spreading.

Municipal water lines have been extended to neighboring residences providing low iron content water that is of much higher overall quality. The overall
consequence is that a reasonable clean-up is being completed that will result in
a usable industrial site being available for development without destroying a
local industry.

11.4   VACUUM-ENHANCED EDUCTOR SYSTEM

A refinery along a shipping canal in the southern United States was plagued
by continuous oil seepage into the canal (Figure 11.12). Local environmental
regulatory authorities were encouraging effective remediation within a limited

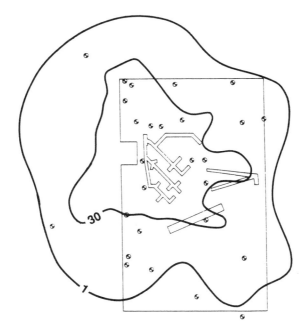

Figure 11.11.   Predicted total organic carbon distribution in shallow groundwater beneath refinery site.

time period. The general geological conditions underlying the site are shown in cross-section A-A´ (Figure 11.13). During construction of the canal, dredged material was pumped onto the bank behind a thin soil dike. After a number of years, industrial facilities, including a refinery, were constructed on this fill area.

Test borings revealed that the fill material consists predominantly of thinly laminated silt with a relatively high clay content. Occasional discontinuous layers of fine sand were encountered at random locations, both horizontally and vertically. The water table gradient under nonpumping conditions was toward the canal, with an approximate gradient of 0.011/ft. Slug tests indicated an effective permeability of 1×10^{-5} cm/s.

The apparent LNAPL thickness observed in monitoring wells was highly variable, ranging from a sheen up to 3 ft accumulation after several days. The spatial distribution of the thickest accumulations suggested a multiple source origin (within the same refinery). Multiple sources were supported by chemical analysis of the product recovered from certain monitoring wells. Each product sample was comprised of an admixture, consisting predominantly of gasoline, kerosene, or crude oil.

Several recovery scenarios were considered for remediation. Initially, con-

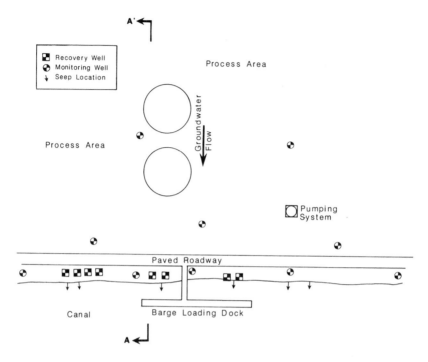

Figure 11.12.   Site facility layout map.

struction of a narrow permeable trench parallel to the canal appeared to be an appropriate interception system. The construction techniques considered was the use of a specially designed deep trenching unit. This type of trench would have included a tile drain leading to a single two-pump recovery well. However, a review of the subsurface site plans and interviews with long-term employees determined that an unknown number of buried pipes traversed the area intended for the trench construction. Disruption of refining operations and the safety considerations resulted in the rejection of this option.

The second option considered was the use of interception wells. One- or two-pump wells could be constructed at calculated spacings to create a hydraulic trough parallel to the canal to intercept the product. This design was considered to be more acceptable to the safety officer and the facility engineer, but was rejected by the maintenance foreman because of the relative complexity of the operation system. The number of submersible pumps and sophisticated electronic controls would have required employment (or training) of technical specialists beyond the cost budgeted under normal operations.

After another review of the practical constraints and technical considerations, it was concluded that an eductor-vacuum enhanced recovery system would be

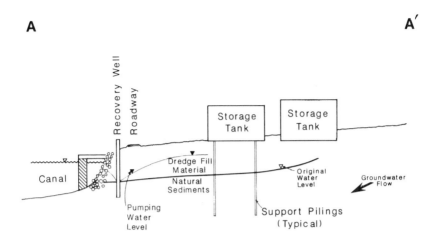

Figure 11.13.   Hydrogeologic cross-section of refinery site.

feasible and would still be cost effective and efficient. Eleven 6-in. diameter wells extending to a depth of 28 ft were installed at strategic locations in the area where product seeps were observed. Each well was serviced by a high-pressure supply and a low-pressure return line. A basic domestic-type deep-well eductor was installed in each well, attached to a drop-pipe, which extended to 25 ft below the surface. A check valve on the drop pipe prevented backflow into the well during service. The top of the casing was sealed with a multiport well top seal to maintain the vacuum. A schematic diagram of a typical recovery well is shown in Figure 11.14.

The pumping unit consisted of a submersible pump installed in a 10,000 gal reservoir tank, which also served as a holding tank (Figure 11.15). Fluids pumped from the wells were all returned to the tank for separation. All the oil and some water produced overflowed from the tank into another oil-water separator and then into the refinery's slop oil treatment system. Clarified water from the bottom of the tank was recycled by the submersible pump through the eductor units to continue operations.

Instrumentation installed on this system was intended to monitor pressure, vacuum, and flow rates, and to prevent accidental spills. Each well casing was equipped with a vacuum gauge. Lines leading to and from the eductor have a site glass to observe flow and are equipped with auxiliary piping for temporary attachment of a blow-rate meter. Pressure was monitored on both high- and low-pressure piping by gauges. Overflow from the tank was measured by a continuous recording flow meter. Oil production was measured in a separate holding tank by periodic withdrawals of oil and water. Level switches in the reservoir tank acted as a safety measure to prevent the levels from becoming too high or too low.

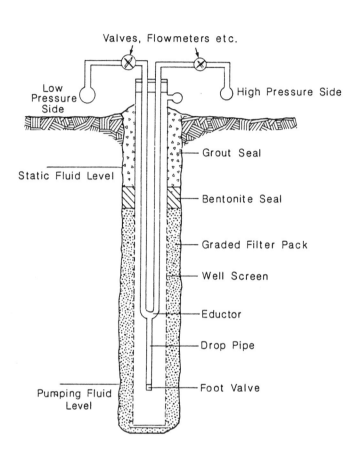

Figure 11.14.   Schematic of typical recovery well construction detail.

Initial production from each of the eleven wells was approximately 1.5 gpm total fluid. Product production from the system was reported to average 15 bpd for the first 60 days and 12 bpd for the next 60 days. Visible seepage to the canal was achieved and the state regulators dismissed further administrative action in exchange for periodic monitoring data.

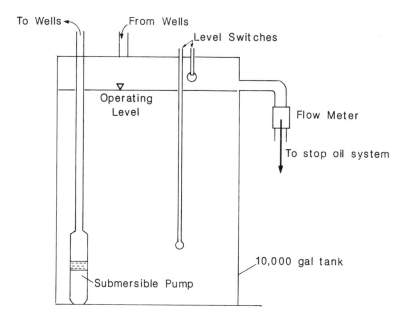

Figure 11.15.   Schematic of reservoir tank.

11.5   COMBINED ONE- AND TWO-PUMP SYSTEM WITH REINJECTION

For the past 6 years, LNAPL occurrence has been investigated beneath an active refinery site in southern California. Numerous monitoring wells, along with LNAPL samples, have been used to evaluate the extent and character of LNAPL occurrence. LNAPL was found to occur as five pools. The main pools each consist of individual accumulations of distinct product types occurring under both perched and water table conditions. Two different recovery and mitigation strategies have been utilized. In relatively high permeability zones, a system of two-pump recovery wells are used to recover fluids; recovered water is reinjected without treatment. In relatively low permeability zones, a system of one-pump recovery wells is used. In the latter case, recovered water is treated prior to disposal.

The property currently occupied by the refinery is a former portion of a large Spanish land grant known as Rancho San Pedro. Geography made the former ranch property a prime location for a refinery being only 3 miles west of Signal Hill, the first rural land north from the shipping facilities of the Los Angeles Harbor and the location of an oil frenzy in the early 1920s.

Over the years, the refinery has produced a range of petroleum products, including liquid petroleum gas, gasoline, chemicals, solvents, distillate fuels, gas oils, lubricating oils, greases, asphalt products, and bunker fuels. The refinery has never produced nor stored pesticides, chlorinated hydrocarbon products, or radioactive materials. The refinery has also never received hazardous wastes from off-site sources during its history. There are currently no unlined storage areas within the refinery. All petroleum liquids have historically been stored in tanks or lined structures.

The primary products of the refinery are currently gasoline, jet fuel, and diesel. Minor products include coke, sulfur, naphtha, and fuel oil. The refinery processes approximately 200,000 barrels of crude oil per day. All crude oil processed is derived from the Alaskan North Slope, although historically some crude was derived from local oil fields via pipelines. This crude is transported to the port by tanker ship and delivered to the refinery by pipeline.

Subsurface geologic and hydrogeologic conditions beneath the facility have been investigated by the drilling and sampling of borings at a minimum of 300 locations, with the subsequent installation of wells(s). These borings extend to a maximum depth of 140 ft. Four types of wells have been installed: monitoring wells, injection wells, recovery wells, and 39 "weep" wells. Monitoring wells screen either the perched or water table zones. The "weep" wells, which have since been abandoned, were screened across both perched and water table zones in an attempt to drain perched LNAPL down to the water table for recovery.

On-site aquifer testing has also been performed in several phases. Typically, pumping tests are performed on newly installed injection and recovery wells to collect the data used to determine optimum pumping or maximum injection rates. Monitoring wells are also tested to determine the location or feasibility of injection and recovery wells.

The characterization of LNAPL in pools underlying the site consisted of collecting and analyzing over 120 LNAPL product samples. LNAPL product samples were distilled and analyzed to determine API gravity, lead, and sulfur content. Boiling point and percent recovery data were used to generate distillation curves.

The site is underlain by a sequence of unconsolidated, stratified, laterally discontinuous deposits of sand, silty sand, clayey silt, and silty clay of Recent and Upper Pleistocene age. A thin veneer of Recent deposits immediately underlies the eastern portion of the site. These deposits are difficult to distinguish from the underlying Upper Pleistocene deposits, which comprise the lower portion of the Lakewood Formation, due to similarities in lithology.

The subsurface soils that comprise the lower portion of the Lakewood Formation can essentially be divided into three informal units, as shown in Figure 11.16. The uppermost unit is comprised predominantly of sand and silty sand with lenses of finer-grained soils consisting of clay, silt, silty clay, clayey silt, and sandy clay. However, a preponderance of finer-grained soils is evident in the northern and eastern portion of the main refinery adjacent to the flood control channel and are, in part, of Recent age. These fine-grained soils reflect

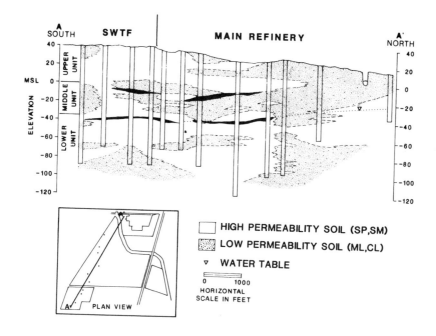

Figure 11.16. Generalized hydrogeologic cross-section showing occurrence of perched and water table LNAPL. Approximate positions of upper, middle, and lower informal units of the Lakewood Formation as shown on the left margin. Vertical exaggeration equals 25:1 (after Testa et al. 1989).

a lateral facies change that is typical of a fluvial depositional environment. Deposits of the uppermost unit extend to depths ranging to about 40 ft below ground surface.

The middle unit is comprised predominantly of fine-grained soils consisting primarily of clay and silt, with subordinate lenses of sand and silty sand. These deposits are typically laterally discontinuous, as is evident in the hydrogeologic cross-section A-A' (Figure 11.16). It is upon these fine-grained layers that perched groundwater occurs beneath the site. These deposits are present at depths ranging from about 35 to 70 ft below ground surface, with a maximum thickness on the order of about 35 ft.

The lower unit is comprised predominantly of coarser grained soils that are similar in bedding and character to those of the uppermost unit described above. These deposits are first encountered at depths of about 50–70 ft below ground surface. All borings drilled were terminated in this unit. Water table LNAPL occurs predominantly within this unit.

Groundwater beneath the site occurs within the Gage aquifer under perched and water table conditions. Beneath the northern half of the main refinery, shallow groundwater appears to occur only under water table conditions. Beneath the southwestern portion and southern half of the main refinery, groundwater occurs under perched conditions upon and within the middle unit. Groundwater is present under water table conditions within the lower unit.

Groundwater encountered under perched conditions beneath the southwestern portion of the site occurs above a laterally discontinuous clay layer which lies at a depth of about 40–50 ft below ground surface (ranging in elevation from about sea level to 12 ft below sea level). The elevation of perched groundwater beneath the main portion of the refinery ranges from 40 to 70 ft below ground surface (ranging in elevation from sea level to about 30 ft below sea level). The majority of the wells completed in the perched aquifer are now dry. Perched conditions in this area appears to occur as individual, discontinuous perching layers.

A water table piezometric surface contour map is presented in Figure 11.17. The elevation of this surface varies from 10 to 40 ft below sea level across the facility. Groundwater occurring under water table conditions is first encountered between 30 and 60 ft in depth. The regional direction of flow of shallow groundwater is toward a depression in the piezometric surface of the Gage aquifer located about 2000 ft to the southwest. Water table flow directions are in general agreement with the regional flow direction, except within the areas of influence of the two recovery systems. Depressions and mounding in the configuration of the water table reflects the continued pumpage and reinjection associated with LNAPL recovery operations.

The series of active recovery wells located in the southeast portion of the main refinery has formed an elongate trough and a closed depression exhibiting about 2 ft of relief. A relatively steep gradient equal to almost 45 ft/mi is present between the flood control channel and recovery wells beneath the eastern portion of this facility.

A limited number of aquifer pumping tests have been conducted within the perched zone. Two separate tests were conducted on a recovery well located in the southwestern portion of the refinery. Low transmissivity values of 100 and 150 gpd/ft were calculated.

Data collected from aquifer pumping tests conducted on LNAPL recovery wells indicated a wide range of transmissivity values within the water table aquifer. Transmissivity values ranged from 2000 to 74,000 gpd/ft, although typical values were between 2500 and 23,000 gpd/ft. The variation in transmissivity and conductivity values reflect the presence of aquifer heterogeneities, including stratification, varying thickness, and partial penetration of the LNAPL recovery wells. Pumping tests conducted on wells screened within the water table aquifer beneath the main refinery indicate that the transmissivity of this material is relatively low, ranging from about 30 to 1620 gpd/ft.

Beneath the facility, LNAPL product occurs as three main pools (pool I, II, and III) and two smaller pools (pool IV and V) of localized occurrence. The

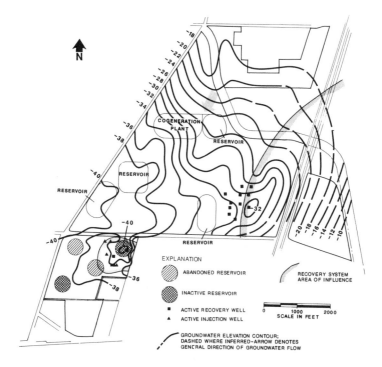

Figure 11.17.   Corrected water table piezometric surface contour map showing regional gradient towards the southwest. Note the impact of recovery systems in their area of influence or capture areas (after Testa et al. 1989).

known extent of pool I, III, IV and V are located entirely beneath the main portion of the refinery. Pool II is located beneath the southwestern portion of the site. LNAPL occurrence beneath the facility is shown in Figure 11.18. For the purposes of this discussion, a pool is defined as an areally continuous accumulation of LNAPL. The three main pools consist of individual accumulations of differing product character and are therefore referred to as coalesced pools. Individual product accumulations within these coalesced pools were delineated on the basis of physical and chemical properties characteristic of free hydrocarbon samples retrieved from the wells. Individual accumulations of relatively uniform product are referred to as subpools, since it is inferred that they coalesced to form areally continuous occurrences or pools. These data suggest multiple sources over time for coalesced pools. The numbering of individual product accumulations is based on product type and areal extent, as summarized in Table 11.1. Data presented include areal extent, groundwater conditions, product types, API gravity range, and lead and sulfur content. Volume determinations for certain pools using the method discussed by Testa and Paczkowski

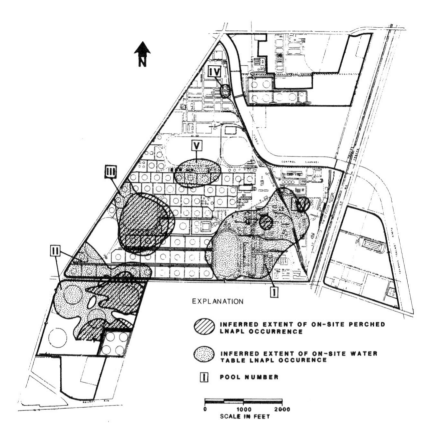

Figure 11.18. **Generalized LNAPL hydrocarbon occurrence map showing individual accumulations that have coalesced to form
pool I, II, and III (after Testa et al. 1989).**

(1989) are presented in Table 11.2. Areal extent is considered a minimum
estimate where pools border the site boundaries and/or extend beyond the limits
of on-site data.

LNAPL recovery at the facility dates back to 1977, when recovery was
initiated at the site by the owner when losses from above-ground storage
structures were discovered. The recovery wells, each utilizing a single positive-
displacement pump, operated intermittently through 1979 until hydrodynamic
conditions caused wells to pump water. About 38,000 barrels were recovered
from the perched zone with this method. In late 1982, a more aggressive
approach was undertaken to delineate the extent of the LNAPL pool and to
design and implement a recovery system. Two 2-pump recovery wells were
installed and put into operation in early 1983. During the latter part of 1983, three
additional two-pump recovery wells were installed. In addition to the two-pump

water table recovery system, a single-pump recovery well was installed in the perched zone in 1984. By the end of 1985, 89 monitoring wells had been installed on site to delineate the free hydrocarbon occurrence and to monitor the effectiveness of the aquifer rehabilitation program.

Exploration for LNAPL beneath the main portion of the refinery commenced in October 1984. Two-in. monitoring wells were initially installed followed by 4-in. test wells in areas of known occurrence. A one-pump recovery system was subsequently designed and installed in the less permeable material encountered beneath this area. The first well was activated in June 1987.

LNAPL recovery operations in the southwestern portion of the refinery have been conducted using a two-pump recovery system. This system currently includes up to 4 two-pump hydrocarbon recovery wells. Each two-pump system uses a 16-in. diameter recovery well, which is designed to accommodate two independently operated pumps placed at different levels within the well. The typical recovery system configuration presently being used is shown in Figure 11.19.

The two pumps within each recovery well are controlled by a series of electrodes that are positioned at predetermined levels within the well. The water pump utilizes a power interrupter probe to detect free hydrocarbon. This probe is positioned above the water intake and is adjusted to automatically turn off the pump when the hydrocarbon interface approaches the pump intake. This prevents the lower pump from accidentally pumping LNAPL to the injection wells.

The upper hydrocarbon pump is switched on and off by a "product" probe, which measures the resistivity of fluids adjacent to the intake of the hydrocarbon pump. When the "product" probe senses the higher resistivity hydrocarbon, it switches on the product pump, which removes the accumulated LNAPL. The water level in the well rises in response to the removal of overlying product. As soon as the water reaches the product probe, the probe shuts down the hydrocarbon pump until the probe again senses that a sufficient quantity of free hydrocarbon has accumulated within the well. The well continues to cycle in this manner as long as free hydrocarbon is entering the well.

The groundwater that is produced by the recovery wells is reinjected into the aquifer from which it was originally pumped through a network of 6- and 2-in. diameter injection wells. The water that is produced at each well is metered, but not consumed or treated in any manner, in accordance with an existing court order.

To evaluate enhanced recovery, a single-pump vacuum-assisted recovery system was employed. This system was temporarily installed to evaluate the recovery of LNAPL from the relatively low-permeability perched zone beneath the northern portion of the site. The system was installed in 1984 and consists of a single 6-in. well. The well was completed with a 10 ft section of screen across the perched zone. The screen was connected to an underlying 30-ft long sump. The annular space from the bottom of the sump to the bottom of the screen was

Table 11.1   Summary of Light Nonaqueous Liquid Hydrocarbon Occurrence

Pool No.	Estimated Extent (acres)	Groundwater Conditions	Subpool/Accumulation No.: Product Type(s)	API Gravity Range	Lead Content Range (g/gal)	Sulfur Content Range (wt. %)
I	2	Perched	Two accumulations both consisting of "light hydrocarbon"[1]	No data	No data	No data
I	82[2]	Water table	I-A: Mixture of gasoline, naphtha & kerosene	53.0–60.6	<0.005–2.00	0.12–0.25
			I-B: Kerosene with smaller amounts of naphtha and light gas oil	33.4–44.2	<0.005–0.02	0.13–0.51
			[3]I-C: Mixture of naphtha and kerosene[3]	52.7	<0.005	0.01
II	25[2]	Perched	II-C: Light gas oil, with smaller amounts of kerosene and heavy gas oil	23.2–27.9	<0.005–0.02	0.09–0.50
			II-D: Light gas oil	20.7–22.2	<0.005	0.90–1.13
II	70[2]	Water table	II-A: Mixture of gasoline naphtha and kerosene	50.7–54.1	0.18–0.62	0.21–0.26
			[3]II-B: Mixture of gasoline, naphtha and kerosene[3]	55.9	0.52	0.11

			Product			
			II-C: Light gas oil, with smaller amounts of kerosene and heavy gas oil	22.8–29.6	<0.005–0.08	0.05–0.53
			II-D: Light gas oil with smaller amount of naphtha and kerosene	20.9–22.9	<0.005	0.95–1.14
III	32	Perched	III-A: Mixture of light and heavy gas oil	19.2–24.0	<0.005	0.92–1.30
III	36[1]	Water table	III-A: Mixture of light and heavy gas oil[3]	22.2	No data	No data
			III-B: Mixture of gasoline and naphtha[3]	42.6	<0.005	0.04
			III-C: Light hydrocarbon(a)[3]	59.1	No data	No data
IV	1	Water table	Unknown	No data	No data	No data
V	17	Water table	Gasoline with smaller amounts of naphtha and kerosene	61.3–69.2	0.12–0.15	0.04–0.51

[1] *Light hydrocarbon* is a field term referring to color and estimated density and is used where no product samples have been analyzed.

[2] Pools border on site boundary or lack well control, resulting in minimum area estimates.

[3] Only one LNAPL sample was analyzed.

Table 11.2   LNAPL Recovery System Effectiveness

LNAPL Pool	Estimated Original Volume Recoverable (barrels)[a]	Recovered to Date (barrels)	Estimated Present Volume Recoverable (barrels)	Estimated Percent Removed
Pool I				
Water table	181,000	28,000	151,000	15
Pool II				
Perched	110,000	2,000	108,000	1.8
Water table	310,000	183,000	127,000	59

[a] 42 gal = 1 barrel.

sealed, allowing for the accumulation of hydrocarbon and water within the sump for recovery, without providing a conduit for downward migration of recoverable product. Another advantage of this construction was that it created a vacuum opposite the area of contamination, instead of pulling in air from zones above and below the LNAPL product layer, which would reduce the overall efficiency of the recovery effort.

A submersible pump was installed within the sump and used to remove the accumulated LNAPL and water from the sump, and to pump it to a storage tank for separation. A vacuum pump created suction, which increased the recovery rate. The effect of vacuum assistance on product recovery rates is illustrated in Figure 11.20.

The main refinery LNAPL recovery system consists of 11 single-pump 4- and 6-in. diameter production wells. Recovery wells are constructed of slotted PVC screens and casing. Four-in. diameter submersible pumps have been installed for the recovery of LNAPL. The submersible pumps installed within the recovery wells are Grundfos pumps of stainless steel construction. These pumps are not adversely affected by water, hydrocarbon, or minor amounts of fine sand and silt produced by the recovery wells.

Due to the low transmissivities of the formations being pumped, the pumps frequently break suction. Current fluid production rates range from as low as 0.4 gpm up to 20 gpm. The total fluid production from the system is approximately 50 gpm. The produced fluid is piped via overhead pipelines to a tank that has been modified to serve as a "gun barrel" oil-water separator. The water is constantly drained from the separator tank through a "water leg". The water leg results in a fairly constant hydrocarbon/water interface level inside the tank, while allowing the water to drain from the tank as it accumulates. The free hydrocarbon is drained off the top of the water, entering a pipe set above the static

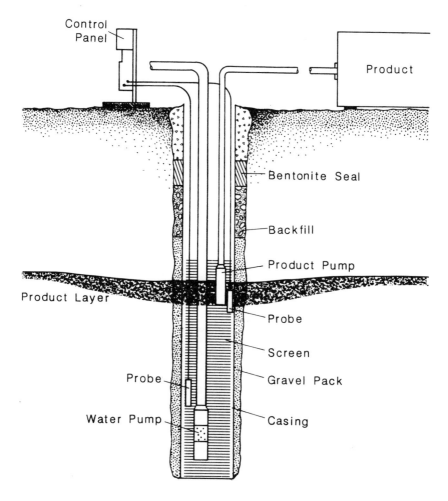

Figure 11.19.   Two-pump system recovery system.

interface. The hydrocarbon product is transported to a storage tank prior to reprocessing. A schematic diagram of the system is presented in Figure 11.21.

The water that is currently being produced by the recovery system is collected and treated on site. The wastewater treatment plant reduces the chemical concentrations to desirable levels and then discharges the treated water into an existing county sanitation district industrial sewer. In this manner, all fluids produced by the recovery system are treated for disposal or are reprocessed.

The LNAPL recovery program in the southwestern portion of the site was initiated in 1982. To date, the volume of LNAPL recovered is approximately 185,000 barrels, which includes about 2,000 barrels from the perched pool and

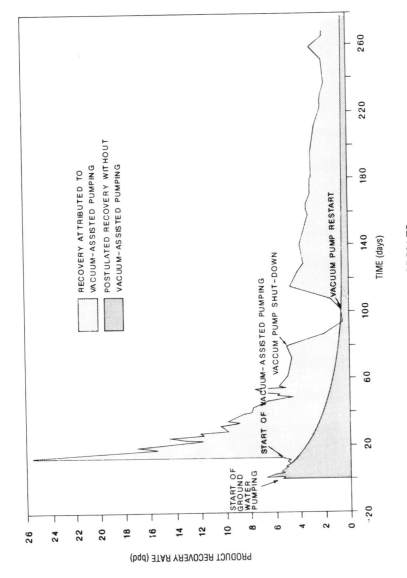

Figure 11.20.   Effectiveness of vacuum assistance on recovery of LNAPL.

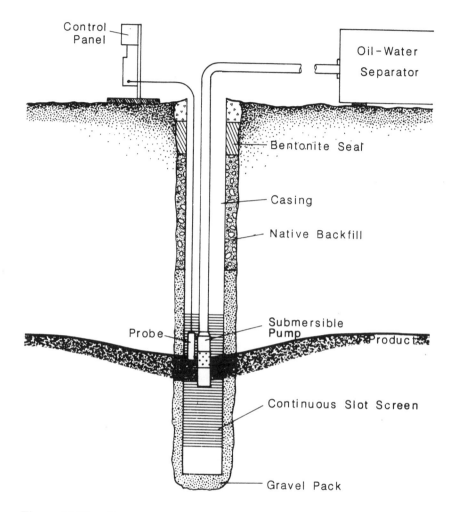

Figure 11.21. One-pump system recovery system.

183,000 barrels from the water table pool. About 28,000 barrels have been recovered to date from beneath the main portion of the refinery.

The two-pump recovery system used at the site utilizes injection wells to

- Create a barrier to off-site migration
- Enhance the hydraulic gradient towards free hydrocarbon recovery wells
- Provide for nonconsumptive water withdrawal

The injection wells at the site have been utilized since 1982 for the disposal of water generated from the hydrocarbon recovery system. The water is reinjected into the aquifer from which it was originally withdrawn so that the quality of the receiving formation is not adversely affected. This type of remediation strategy also allows less strain on the facility's wastewater handling capabilities.

The injection well system has been expanded in conjunction with the operation of the recovery system. Due to operational constraints and the locations of injection wells in service, injection rates and volumes have varied. As of January 1989, approximately 1330 acre-ft of groundwater had been pumped and reinjected. Pumpage of this water coincided with the recovery of about 183,000 barrels of free hydrocarbon. Fryberger and Shepard (1987) summarized the benefits of reinjection without treatment of the generated water during LNAPL aquifer remediation programs.

Making some rough approximations of recovery without two-pump systems, Testa and Paczkowski (1989) gave an indication of the magnitude of the injection system impact. From experience at other sites with similar subsurface hydrogeologic and geologic conditions and LNAPL occurrence, and where recovery has been attempted without the creation of a cone of depression, single-pump recovery systems (typically referred to as "skimming" or "pneumatic" recovery systems) have the following recovery performance histories. The initial recovery rates from 6-in. diameter recovery wells may be about 8–15 bpd. After about 1–3 months, the long-term sustained recovery will drop to less than five bpd. After 1 year of operation recovery will be about 1 bpd or less. The radius of influence of these wells is minimal and is difficult to identify.

From six such recovery wells, the volume of LNAPL recovered in 5 years might have been

$$
\begin{array}{llll}
6 \text{ wells} \times 15 \text{ bpd} \times & 90 \text{ d} & = & 8.100 \text{ barrels} \\
6 \text{ wells} \times 5 \text{ bpd} \times & 270 \text{ d} & = & 8.100 \text{ barrels} \\
6 \text{ wells} \times 1 \text{ bpd} \times & 4 \text{ years} & = & \underline{8.760 \text{ barrels}} \\
& & & 24.960 \text{ barrels}
\end{array}
$$

This would be about 159,000 barrels less than that actually recovered to date. Assuming that one skimming well would recover as much as 4200 barrels in 5 years (24,960 barrels/6 wells), then 44 such recovery wells would have been required to achieve the same volume recovered.

Although this is a simplistic analysis, it points out very clearly that recovery using two-pump systems is significantly more effective than by skimming alone. This conclusion is supported by data recorded for the two respective recovery systems. The average recovery rates over the first year of production range from 50 to 150 bpd of LNAPL and from 350 to 3000 bpd of water per two-pump recovery well. In comparison, one-pump recovery wells have averaged approximately 5 bpd of LNAPL and 140 bpd of water over the first year of production. The two-pump recovery system does have disadvantages, including increased operation and maintenance costs, and plugging of injection wells, reflecting an increase in microbial activity. Future endeavors will concentrate on enhanced recovery of LNAPL, delineation, and subsequent aquifer rehabilitation of the dissolved hydrocarbon plume.

REFERENCES

1. Fryberger, J. S. and Shepard, D. C., 1987, Reinjection of water at hydrocarbon recovery sites: International Symposium on Class V Injection Well Technology, U.S. Environmental Protection Agency and Underground Injection Practices Council, 19 p.
2. Henry, E. C. and Hayes, D., Practical Approach to Hydrocarbon Recovery at Marine Terminals: In Proceedings of the National Water Well Association and American Petroleum Institute Conference on Petroleum Hydrocarbons and Organic Chemicals in Groundwater: Prevention, Detection and Restoration, November, 1987, p. 75-90.
3. Henry, E. C., Testa, S. M., Hayes, D., and Hodder, E. H., 1988, Advantages of Suction Lift Hydrocarbon Recovery Systems: Application at Three Hydrogeologic Environments in California: Proceedings of the National Water Well Association FOCUS Conference on Southwestern Groundwater Issues, March, 1988, p. 487-502.
4. Testa, S. M., Henry, E. C., and Hayes, D. , 1988, Impact of the Newport-Inglewood structural zone on hydrogeologic mitigation efforts - Los Angeles basin, California: Proceedings of the National Water Well Association FOCUS Conference on Southwestern Groundwater Issues, p.181-203.
5. Testa, S. M., Baker, D., and Avery, P., 1989, Field Studies in Occurrence, Recoverability, and Mitigation Strategy for Free Phase Liquid Hydrocarbon: In Environmental Concerns in the Petroleum Industry (Edited by Testa, S.M.), Pacific Section of the American Association of Petroleum Geologists Symposium Volume, p. 57-81.
6. Testa, S. M. and M. Paczkowski, 1989, Volume Determination and Recoverability of Free Hydrocarbon: *Ground Water Monitoring Review*, Winter Issue, p. 120-128.

APPENDIX A
SPECIFIC GRAVITY
CORRESPONDING TO
API GRAVITY

Appendix A   Specific Gravity Corresponding to API Gravity

API Gravity	Decimal Parts of a Degree									
	0.0	0.1	0.2	0.3	0.4	0.5	0.6	0.7	0.8	0.9
10	1.0000	0.9993	0.9986	0.9979	0.9972	0.9965	0.9958	0.9951	0.9944	0.9937
11	0.9930	0.9923	0.9916	0.9909	0.9902	0.9895	0.9888	0.9881	0.9874	0.9868
12	0.9861	0.9854	0.9847	0.9840	0.9833	0.9826	0.9820	0.9813	0.9806	0.9799
13	0.9792	0.9786	0.9779	0.9772	0.9765	0.9759	0.9752	0.9745	0.9738	0.9732
14	0.9725	0.9718	0.9712	0.9705	0.9698	0.9692	0.9685	0.9679	0.9672	0.9665
15	0.9659	0.9652	0.9646	0.9639	0.9632	0.9626	0.9619	0.9613	0.9606	0.9600
16	0.9593	0.9587	0.9580	0.9574	0.9567	0.9561	0.9554	0.9548	0.9541	0.9535
17	0.9529	0.9522	0.9516	0.9509	0.9503	0.9497	0.9490	0.9484	0.9478	0.9471
18	0.9465	0.9459	0.9452	0.9446	0.9440	0.9433	0.9427	0.9421	0.9415	0.9408
19	0.9402	0.9396	0.9390	0.9383	0.9377	0.9371	0.9365	0.9358	0.9352	0.9346
20	0.9340	0.9334	0.9328	0.9321	0.9315	0.9309	0.9303	0.9297	0.9291	0.9285
21	0.9279	0.9273	0.9267	0.9260	0.9254	0.9248	0.9242	0.9236	0.9230	0.9224
22	0.9218	0.9212	0.9206	0.9200	0.9194	0.9188	0.9182	0.9176	0.9170	0.9165
23	0.9159	0.9153	0.9147	0.9141	0.9135	0.9129	0.9123	0.9117	0.9111	0.9106
24	0.9100	0.9094	0.9088	0.9082	0.9076	0.9071	0.9065	0.9059	0.9053	0.9047
25	0.9042	0.9036	0.9030	0.9024	0.9018	0.9013	0.9007	0.9001	0.8996	0.8990
26	0.8984	0.8978	0.8973	0.8967	0.8961	0.8956	0.8950	0.8944	0.8939	0.8933
27	0.8927	0.8922	0.8916	0.8911	0.8905	0.8899	0.8894	0.8888	0.8883	0.8877
28	0.8871	0.8866	0.8860	0.8855	0.8849	0.8844	0.8838	0.8833	0.8827	0.8822
29	0.8816	0.8811	0.8805	0.8800	0.8794	0.8789	0.8783	0.8778	0.8772	0.8767

30	0.8713	0.8718	0.8724	0.8729	0.8735	0.8740	0.8745	0.8751	0.8756	0.8762
31	0.8660	0.8665	0.8670	0.8676	0.8681	0.8686	0.8692	0.8697	0.8702	0.8708
32	0.8607	0.8612	0.8618	0.8623	0.8628	0.8633	0.8639	0.8644	0.8649	0.8654
33	0.8555	0.8560	0.8565	0.8571	0.8576	0.8581	0.8586	0.8591	0.8597	0.8602
34	0.8504	0.8509	0.8514	0.8519	0.8524	0.8529	0.8534	0.8540	0.8545	0.8550
35	0.8453	0.8458	0.8463	0.8468	0.8473	0.8478	0.8483	0.8488	0.8493	0.8498
36	0.8403	0.8408	0.8413	0.8418	0.8423	0.8428	0.8433	0.8438	0.8443	0.8448
37	0.8353	0.8358	0.8363	0.8368	0.8373	0.8378	0.8383	0.8388	0.8393	0.8398
38	0.8304	0.8309	0.8314	0.8319	0.8324	0.8328	0.8333	0.8338	0.8343	0.8348
39	0.8256	0.8260	0.8265	0.8270	0.8275	0.8280	0.8285	0.8289	0.8294	0.8299
40	0.8208	0.8212	0.8217	0.8222	0.8227	0.8232	0.8236	0.8241	0.8246	0.8251
41	0.8160	0.8165	0.8170	0.8174	0.8179	0.8184	0.8189	0.8193	0.8198	0.8203
42	0.8114	0.8118	0.8123	0.8128	0.8132	0.8137	0.8142	0.8146	0.8151	0.8156
43	0.8067	0.8072	0.8076	0.8081	0.8086	0.8090	0.8095	0.8100	0.8104	0.8109
44	0.8022	0.8026	0.8031	0.8035	0.8040	0.8044	0.8049	0.8054	0.8058	0.8063
45	0.7976	0.7981	0.7985	0.7990	0.7994	0.7999	0.8003	0.8008	0.8012	0.8017
46	0.7932	0.7936	0.7941	0.7945	0.7949	0.7954	0.7958	0.7963	0.7967	0.7972
47	0.7887	0.7892	0.7896	0.7901	0.7905	0.7909	0.7914	0.7918	0.7923	0.7927
48	0.7844	0.7848	0.7852	0.7857	0.7861	0.7865	0.7870	0.7874	0.7879	0.7883
49	0.7800	0.7805	0.7809	0.7813	0.7818	0.7822	0.7826	0.7831	0.7835	0.7839
50	0.7758	0.7762	0.7766	0.7770	0.7775	0.7779	0.7783	0.7788	0.7792	0.7796
51	0.7715	0.7720	0.7724	0.7728	0.7732	0.7736	0.7741	0.7745	0.7749	0.7753
52	0.7674	0.7678	0.7682	0.7686	0.7690	0.7694	0.7699	0.7703	0.7707	0.7711
53	0.7632	0.7636	0.7640	0.7645	0.7649	0.7653	0.7657	0.7661	0.7665	0.7669
54	0.7591	0.7595	0.7599	0.7603	0.7608	0.7612	0.7616	0.7620	0.7624	0.7628
55	0.7551	0.7555	0.7559	0.7563	0.7567	0.7571	0.7575	0.7579	0.7583	0.7587

Appendix A   Specific Gravity Corresponding to API Gravity (continued)

API Gravity	Decimal Parts of a Degree									
	0.0	0.1	0.2	0.3	0.4	0.5	0.6	0.7	0.8	0.9
56	0.7547	0.7543	0.7539	0.7535	0.7531	0.7527	0.7523	0.7519	0.7515	0.7511
57	0.7507	0.7503	0.7499	0.7495	0.7491	0.7487	0.7483	0.7479	0.7475	0.7471
58	0.7467	0.7463	0.7459	0.7455	0.7451	0.7447	0.7443	0.7440	0.7436	0.7432
59	0.7428	0.7424	0.7420	0.7416	0.7412	0.7408	0.7405	0.7401	0.7397	0.7393
60	0.7389	0.7385	0.7381	0.7377	0.7374	0.7370	0.7366	0.7362	0.7358	0.7354
61	0.7351	0.7347	0.7343	0.7339	0.7335	0.7332	0.7328	0.7324	0.7320	0.7316
62	0.7313	0.7309	0.7305	0.7301	0.7298	0.7294	0.7290	0.7286	0.7283	0.7279
63	0.7275	0.7271	0.7268	0.7264	0.7260	0.7256	0.7253	0.7249	0.7245	0.7242
64	0.7238	0.7234	0.7230	0.7227	0.7223	0.7219	0.7216	0.7212	0.7208	0.7205
65	0.7201	0.7197	0.7194	0.7190	0.7186	0.7183	0.7179	0.7175	0.7172	0.7168
66	0.7165	0.7161	0.7157	0.7154	0.7150	0.7146	0.7143	0.7139	0.7136	0.7132
67	0.7128	0.7125	0.7121	0.7118	0.7114	0.7111	0.7107	0.7103	0.7100	0.7096
68	0.7093	0.7089	0.7086	0.7082	0.7079	0.7075	0.7071	0.7068	0.7064	0.7061
69	0.7057	0.7054	0.7050	0.7047	0.7043	0.7040	0.7036	0.7033	0.7029	0.7026
70	0.7022	0.7019	0.7015	0.7012	0.7008	0.7005	0.7001	0.6998	0.6995	0.6991

The specific gravity corresponding to an API gravity of 32.7, for example, will be found on line "32" in column "0.7".

API Gravity and Corresponding Weights and Pressure at 60°F

API Gravity	Weight lb/gal	Weight lb/bbl	Weight lb/cu ft	psi/ft	ft/psi
10	8.328	349.8	62.37	0.4331	2.309
15	8.044	337.9	60.24	0.4183	2.391
20	7.778	326.7	58.25	0.4045	2.472
25	7.529	316.2	56.39	0.3916	2.554
26	7.481	314.2	56.03	0.3891	2.570
27	7.434	312.2	55.68	0.3866	2.587
28	7.387	310.3	55.33	0.3842	2.603
29	7.341	308.3	54.99	0.3818	2.619
30	7.296	306.4	54.65	0.3795	2.635
31	7.251	304.5	54.31	0.3771	2.652
32	7.206	302.7	53.97	0.3748	2.668
33	7.163	300.9	53.65	0.3726	2.684
34	7.119	298.8	53.33	0.3703	2.701
35	7.076	297.2	53.00	0.3860	2.717
36	7.034	295.4	52.69	0.3650	2.733
37	6.993	293.7	52.38	0.3637	2.750
38	6.951	291.9	52.07	0.3616	2.765
39	6.910	290.2	51.76	0.3594	2.782
40	6.870	288.6	51.46	0.3574	2.798
41	6.830	286.9	51.16	0.3553	2.815
42	6.790	285.2	50.86	0.3532	2.831

API Gravity and Corresponding Weights and Pressure at 60°F (continued)

API Gravity	Weight lb/gal	Weight lb/bbl	Weight lb/cu ft	psi/ft	ft/psi
43	6.752	283.6	50.58	0.3512	2.847
44	6.713	282.0	50.29	0.3492	2.864
45	6.675	280.4	50.00	0.3472	2.880
46	6.637	278.8	49.72	0.3453	2.896
47	6.600	277.3	49.44	0.3433	2.913
48	6.563	275.7	49.17	0.3414	2.929
49	6.526	274.2	48.89	0.3395	2.945
50	6.490	272.7	48.62	0.3376	2.962
51	6.455	271.2	48.36	0.3358	2.978
52	6.420	269.7	48.09	0.3340	2.994
53	6.385	268.2	47.83	0.3321	3.011
54	6.350	266.7	47.58	0.3304	3.029
55	6.316	265.4	47.32	0.3286	3.043
56	6.283	264.0	47.07	0.3269	3.059
57	6.249	262.6	46.82	0.3251	3.076
58	6.216	261.2	46.57	0.3234	3.092
59	6.184	259.8	46.33	0.3217	3.108
60	6.151	258.3	46.09	0.3200	3.125
61	6.119	257.1	45.85	0.3184	3.141
62	6.087	255.8	45.61	0.3167	3.158
63	6.056	254.5	45.37	0.3151	3.174

64	6.025	253.2	45.14	0.3135	3.190
65	5.994	251.9	44.91	0.3119	3.206
70	5.845	245.6	43.80	0.3041	3.288
75	5.703	239.7	42.74	0.2968	3.369
80	5.568	234.0	41.73	0.2897	3.452
85	5.440	228.6	40.76	0.2831	3.533
90	5.316	223.4	39.84	0.2767	3.614
95	5.199	218.5	38.96	0.2706	3.695
100	5.086	213.8	38.12	0.2647	3.778

APPENDIX B
VISCOSITY CONVERSION TABLE

Appendix B **Viscosity Conversion Table**

SSU Seconds Saybolt Universal	SSF Seconds Saybolt Furel	Kinematic Viscosity Centistokes (centistokes)	Seconds Redwood (standard)
31	—	1.00	29
35	—	2.56	32.1
40	—	4.30	36.2
50	—	7.40	44.3
60	—	10.20	52.3
70	12.95	12.83	60.9
80	13.70	15.35	69.2
90	14.44	17.80	77.6
100	15.24	20.20	85.6
150	19.30	31.80	128
200	23.5	43.10	170
250	28.0	54.30	212
300	32.5	65.40	254
400	41.9	87.60	338
500	51.6	110.0	423
600	61.4	132	508
700	71.1	154	592
800	81.0	176	677
900	91.0	198	762
1000	100.7	220	896
1500	150	330	1270
2000	200	440	1690
2500	250	550	2120
3000	300	660	2540
4000	400	880	3380
5000	500	1100	4230
6000	600	1320	5080
7000	700	1540	5920
8000	800	1760	6770
9000	900	1980	7620
10000	1000	2200	8460
15000	1500	3300	13700
20000	2000	4400	18400

Source: A.F.L. Industrise Catalog, West Chicago, IL.

Appendix B Viscosity Conversion Table (continued)

$$\frac{\text{Kinematic}}{\text{viscosity}} = \frac{\text{Absolute visc.}}{\text{Specific gravity}}$$
$$\text{(Stokes)}$$

$$1 \text{ Centistoke} = \frac{\text{Stoke}}{100}$$

$$1 \text{ Centipoise} = \frac{\text{Poise}}{100}$$

1 Stoke   = 100 centistokes

1 Poise   = 100 centipoises

Notes: (a) The term *centipoises* is referred to commonly as a measure of kinematic viscosity. Convert centipoises to centistokes by dividing by the specific gravity of the solution at the operating temperature.
(b) Plotting viscosity. If viscosity is known at any two temperatures, the viscosity at other temperatures can be obtained by plotting the viscosity against temperature in °F on log paper. The points lie in a straight line.

APPENDIX C
VISCOSITY AND SPECIFIC GRAVITY OF COMMON PETROLEUM PRODUCTS

Appendix C   Viscosity and Specific Gravity of Common Petroleum Products

Liquid	Specific Gravity	40°F	60°F	70°F	80°F	100°F	120°F	130°F	140°F	160°F
Crankcase Oils - Automobile Lubricating Oils										
SAE 10	.88–.935	1500–2400	600–900	—	300–400	170–220	110–330	—	75–90	60–65
SAE 20	.88–.935	2400–9000	900–3000	—	400–1100	220–550	130–280	—	90–170	65–110
SAE 30	.88–.935	9000–14000	3000–4400	—	1100–1800	550–800	280–400	—	170–240	110–150
SAE 40	.88–.935	14000–19000	4400–6000	—	1800–2400	800–1100	400–550	—	240–320	150–200
SAE 50	.88–.935	19000–45000	6000–10000	—	2400–4000	1100–1800	550–850	—	320–480	200–280
SAE 60	.88–.935	45000–60000	10000–17000	—	4000–6000	1800–2500	850–1200	—	480–580	280–380
SAE 70	.88–.935	60000–120000	17000–45000	—	6000–10000	2500–4000	1200–1800	—	580–900	380–500
Transmission Oils - Automobile Transmission Gear Lubricants										
SAE 90	.88–.935	14000	5500	—	2200	1100	650	—	380	240
SAE 140	.88–.935	35000	12000	—	5000	2200	1200	—	650	400
SAE 250	.88–.935	160000	50000	—	18000	7000	3300	—	1700	1000
Tars										
Coke oven-tar	1.12+	—	—	3000–8000	—	650–1400	—	—	—	—
Gas house-tar	1.16–1.3	—	—	15000–300000	—	2000–20000	—	—	—	—
Crude Oils										
Texas, Oklahoma	.81–.916	—	—	100–700	—	34–210	—	—	—	—
Wyoming, Montana	.86–.88	—	—	100–1100	—	46–320	—	—	—	—

	Sp. gr.									
California	.78–92	—	—	100–4500	—	34–700	—	—	—	—
Pennsylvania	.8–85	—	—	100–200	—	38–86	—	—	—	—
Water	1.0	31.5	31.5	—	31.5	31.5	31.5	—	31.5	31.5
Gasoline	.68–74	30	30	—	30	30	30	—	30	30
Jet Fuel	.74–85	35	35	—	35	35	35	—	35	35
Kerosene	.78–82	42	38	—	34	33	31	—	30	30
<u>Fuel Oil and Diesel Oil</u>										
No.1 fuel oil	.82–95	40	38	—	35	33	31	—	30	30
No. 2 fuel oil	.82–95	70	50	—	45	40	—	—	—	—
No. 3 fuel oil	.82–95	90	68	—	53	45	40	—	—	—
No. 5A fuel oil	.82–95	1000	400	—	200	100	75	—	60	40
No. 5B fuel oil	.82–95	1300	600	—	490	400	330	—	290	240
No. 6 fuel oil	.82–95	—	70000	—	20000	90000	1900	—	900	500
No. 2D diesel fuel oil	.82–95	100	68	—	53	45	40	—	36	35
No. 3D diesel fuel oil	.82–95	200	120	—	80	60	50	—	44	40
No. 4D diesel fuel oil	.82–95	1600	600	—	280	140	90	—	68	54

Appendix C Viscosity and Specific Gravity of Common Petroleum Products (continued)

Liquid	Specific Gravity	40°F	60°F	70°F	80°F	100°F	120°F	130°F	140°F	160°F
No. 5D diesel fuel oil	.82–.95	15000	5000	—	2000	900	400	—	260	160
Navy No. 1 fuel oil	.989	4000	1100	—	600	380	200	—	170	90
Navy No. 2 fuel oil	1.0	—	24000	—	8700	3500	1500	—	900	480
Gas	.887	180	90		60	50	45		—	40
Insulating	—	350	150		90	65	50		45	—
Lard	.912–.925	1100	600		380	287	180		140	90
Linseed	.925–.939	1500	500		250	143	110		85	70
Raw										
Menhadden	.933	1500	500	—	250	140	110	—	80	70
Neats foot	.917	—	1000	—	430	230	160	—	100	80
Olive	.912–.918	1500	550	—	320	200	150	—	100	80
Palm	.924	1700	700	—	380	221	160	—	120	90
Peanut	.920	1200	500	—	300	195	150	—	100	80
Quenching	—	2400	900	—	450	250	180	—	130	90
Rape seed	.919	2400	900	—	450	250	180	—	130	90
Rosin	.980	28000	7800	—	3200	1500	900	—	500	300
Rosin (wood)	1.09	Extremely viscose								
Sesame	.923	1100	500	—	290	184	130	—	90	60
Soya bean	.927–.98	1200	475	—	270	165	120	—	80	70
Sperm	.883	360	250	—	170	110	90	—	70	60

Turbine (light)	.91	500	350	—	230	150	—	—	—	—
Turbine (heavy)	.91	3000	1400	—	700	330	200	—	150	100
Whale	.925	900	450	—	275	170	140	—	100	80

Source: A.F.L. Industries Catalog, West Chicago, IL.

INDEX